Gene Mazzola
NMR Facility
Dept. of Chemistry
Univ. of Maryland

NMR OF CHEMICALLY EXCHANGING SYSTEMS

NMR OF CHEMICALLY EXCHANGING SYSTEMS

Jerome I. Kaplan

Krannert Institute of Cardiology,
Physics Department, IUPUI, and
Indianapolis Center for Advanced Research
Indianapolis, Indiana

Gideon Fraenkel

Department of Chemistry
The Ohio State University
Columbus, Ohio

1980

ACADEMIC PRESS

A Subsidiary of Harcourt Brace Jovanovich, Publishers

New York London Toronto Sydney San Francisco

ACADEMIC PRESS, INC.
111 Fifth Avenue, New York, New York 10003

United Kingdom Edition published by
ACADEMIC PRESS, INC. (LONDON) LTD.
24/28 Oval Road, London NW1 7DX

Library of Congress Cataloging in Publication Data

Kaplan, Jerome I
Nmr of chemically exchanging systems.

Includes bibliographical references and index.
1. Nuclear magnetic resonance spectroscopy.
2. Chemical reaction, Rate of. I. Fraenkel, Gideon, joint author. II. Title.
QD96.N8K36 541'.39 79-50217
ISBN 0-12-397550-6

PRINTED IN THE UNITED STATES OF AMERICA

80 81 82 83 9 8 7 6 5 4 3 2 1

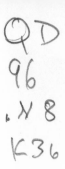

CONTENTS

PREFACE

The purpose of this book is to present, in one place, a unified density matrix formalism for calculating NMR lineshapes for all kinds of exchanging systems. The treatment is illustrated for a variety of conditions including high rf power[1,2] and double resonance,[3–5] and is applicable to both liquids and liquid crystals. Account is also taken of individual nuclear relaxation processes. This book has been written in such a way that a scientist familiar with the salient features of NMR spectroscopy (as described in Ref. 6) can learn how to obtain kinetic information via lineshape analysis.

It is assumed that the reader is familiar with NMR of static molecules (slow exchange) and thus that he understands the nuclear spin Hamiltonian

$$\mathcal{H} = \sum_s \omega_{0s} I_s^z + \sum_{s>t} J_{s,t} I_s \cdot I_t$$

its eigenvalues, eigenstates, and the rf induced transitions between its eigenstates.[6] A brief review to serve only as a reminder of spin operators is given in the Appendix to Chapter III.

All calculations will be derived using the density matrix equation, the subject of Chapter II. No previous knowledge of the density matrix is needed.

The outline of this book is as follows: A derivation of the absorption due to a single spin using the classical Bloch equation as a means for introducing some of the concepts involved in obtaining the absorption lineshape is presented in Chapter I. A derivation of the total (totally isolated) density matrix equation is given in Chapter II. In Chapter III we obtain the density matrix equation for a single spin type—the same problem solved with the Bloch equation of Chapter I. General concepts involved using the density matrix equation are emphasized. Then in Chapter IV we show how to reduce the global density matrix equation to that of the spin density matrix equation. The interaction of the spin density matrix with the "outside world" is made in terms of the relaxation operator R. The development of an exchange operator E that allows for chemical exchange is discussed in Chapter V. A derivation of the absorption of a general exchanging system in the presence of a low rf field is presented in Chapter VI. Density matrix equations are derived for several typical exchanging systems to illustrate procedures.

In Chapter VII we consider the density matrix equation for a general exchanging system at high rf power and show how one has to solve for all elements of the density matrix. These results are utilized in Chapter VIII for calculating the NMR absorption of exchanging systems at two rf fields, one large and one small. Included there are some comments on effects often seen in double resonance. Finally in Chapter IX there are some remarks on transients as applied to chemically exchanging systems.

At this point we would like to issue a caveat about Chapter IV. It is the most difficult chapter in the book. The principal results of this chapter, which are used throughout the book, are summarized in Section 3 of Chapter IV. Readers who only wish to learn the kinetic techniques can safely skip Chapter IV altogether. On the other hand the serious student of NMR should definitely work through this chapter.

While this work is not meant to be a comprehensive review, there are references leading to original sources as well as some for supplementary reading.

At the time of this writing it appears that some of the material discussed in this book represents areas in which little or no work has been done so far. We hope these sections will stimulate new experimental work and point the way to how problems can be handled in the future.

Our original interest in NMR lineshape analysis comes from contacts with J. D. Roberts, California Institute of Technology, and with Saul Meiboom and his group at the Weizmann Institute of Science.

We express our thanks for computer programming to Dr. Alice Fraenkel, Department of Chemistry, The Ohio State University, and Dr. Aaron Supowit, Instruction and Research Computer Center, The Ohio State University. Also, we appreciate the hospitality extended by Dr. Robert E. Carter, Department of Organic Chemistry, Lund Institute of Technology, during the time the authors spent in Lund.

Finally, thanks are due to the Office of Naval Research and the National Science Foundation for supporting some of our work.

REFERENCES

1. R. K. Harris and N. C. Pyper, *Mol. Phys.* **21**, 467 (1971); R. K. Harris and K. M. Worvill, *J. Magn. Reson.*, **9**, 294, 383 (1973).
2. J. Kaplan, P. P. Yang, and G. Fraenkel, *J. Chem. Phys.* **60**, 4840 (1974).
3. P. P. Yang and S. L. Gordon, *J. Chem. Phys.* **54**, 1779 (1971); B. M. Fung and P. M. Olympia, *Mol. Phys.* **19**, 685 (1970).
4. J. M. Anderson, *J. Magn. Reson.* **4**, 184 (1971).
5. J. I. Kaplan, P. P. Yang, and G. Fraenkel, *J. Amer. Chem. Soc.* **97**, 3881 (1975).
6. R. Lynden-Bell and R. K. Harris, "Nuclear Magnetic Resonance Spectroscopy." Appleton, New York, 1969.

NMR OF CHEMICALLY EXCHANGING SYSTEMS

Chapter I
INTRODUCTION

1. Historical Introduction

The contribution of molecular motion to nuclear magnetic relaxation was first discussed by Bloembergen *et al.* in 1948.[1] One specific kind of motion is the structural rearrangement which attends an internal molecular rotation. An example of such nuclear relaxation was uncovered by Gutowsky and Pake[2] who correlated the temperature dependence of the proton NMR lineshapes of solid 1,1,1-trichloroethane with internal rotation, about the C—C bond. Similar early studies on calcium sulfate[3] and ammonium chloride[4] provided information on rotation of ions in solid lattices. In these cases thermally induced motions have the effect of averaging the intramolecular dipole–dipole couplings as well as the dipole–dipole relaxation with the result that the linewidths in these solids decrease with increasing temperature.

The idea that exchange processes might have an effect on the NMR spectrum was first proposed by Liddel and Ramsey[5] in connection with results of NMR studies on ethanol reported by Arnold.[6] Gutowsky and Saika[7] treated the problem theoretically with modified Bloch equations. This procedure was then applied to the problem of hindered rotation in liquid *N,N*-dimethylformamide,[8] **1**. Conjugation within the amide moiety (**2**, **3**) renders the molecule planar and the methyls magnetically nonequivalent. So below $-20°$ the methyl resonance consists of a doublet. Rotation about the C_0—N bond exchanges the methyl environments. When the rotation rate is comparable to the inverse nuclear relaxation time there is uncertainty in the methyl absorption frequencies and the two lines broaden. At rates near the resonance frequency shift between the methyls there is considerable coalescence and overlap. With further increases in rate the methyl resonances become progressively averaged out and one line is seen at 120°.

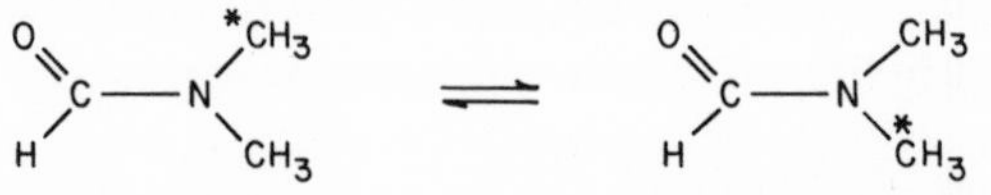

1

2 3

It is because NMR lines of liquids are so narrow (0.05 to 2-Hz wide) compared to their frequency separations and correspondingly relaxation times so long that rate processes with pseudo-first-order rate constants of 10^{-1} to 10^{6} sec^{-1} (it depends on the system) have such a profound effect on the NMR lineshapes. Roberts has compared the perturbation of relaxation by exchange effects to photographing a moving object using a camera with a slow shutter; the result is a blurred image![9]

Cold methanol at $-20°$ shows a 4-Hz methyl, hydroxyl proton coupling constant. At higher temperatures, as well as in the presence of acids and bases, OH proton exchange processes have the effect of averaging this coupling constant.[10] When an OH proton on a methanol molecule is replaced by a second one the latter has a 50% probability of having the same spin state as the first proton, so half the exchanges contribute to the line averaging. Note that if the original proton just ionizes off and comes straight back, as in the reaction

$$CH_3OH \rightleftharpoons CH_3O^- + H^+, \tag{1-1}$$

then there is no effect on the spectrum because the proton spin state does not have time to change! In this case line averaging is the result of a bimolecular mechanism.

The effects just described were first detected by Holm and Maxwell[11] using spin-echo measurements. Later Grunwald and Meiboom carried out lineshape analysis on this system and uncovered the steps responsible for

OH exchange[10]:

$$CH_3O^*H + CH_3OH^* \rightleftharpoons CH_3O^{*-} + CH_3OHH^{*+}, \quad (1\text{-}2)$$

$$CH_3O^*H + CH_3OH^* \rightleftharpoons CH_3O^*H^* + CH_3OH, \quad (1\text{-}3)$$

$$CH_3OH_2^+ + CH_3OH^* \rightleftharpoons CH_3OHH^{*+} + CH_3OH, \quad (1\text{-}4)$$

$$CH_3O^*H + CH_3O^- \rightleftharpoons CH_3O^{*-} + CH_3OH. \quad (1\text{-}5)$$

Among examples of early work on exchange using NMR are the H—F bond exchange in hydrogen fluoride and related systems by Solomon and Bloembergen[12] and the calculation of exchange rates in ethanol by Arnold.[13]

Ogg reported the first example of exchange in a system containing a quadrupolar nucleus.[14] He observed coalescence and averaging with increasing temperature of the aluminum hydrogen coupling in the system $Al_2B_4H_{18} + Al(BH_4)_3$.

Through the progression of developments since 1948—from the discussion of lattice motion effects on relaxation to molecular motion, hindered rotation to consideration of effects due to all kinds of chemical reorganization—it has come to be recognized that the NMR lineshapes of a large variety of systems undergoing different kinds of exchange processes depend on the rates of some exchange processes (see, e.g., Ref. 15), in effect a contribution to relaxation.

The umbrella term *chemical reorganization* covers the innumerable ways in which chemical species can undergo exchange processes—from rotation,[16] pseudorotation,[17] ring inversion,[18] configurational inversion,[19] and other intramolecular rearrangements to bimolecular chemical exchange. A few examples of the latter include proton transfer (see, e.g., Ref. 20), carbon metal bond exchange (reviewed in Ref. 21), and metal ligand coordination exchange.[22] It is important to be aware that all these processes have been studied as *equilibrium systems*. The concentrations of the constituents of these samples do not vary as a function of time—but the exchange processes take place all the time forward and backward at the same rate. In fact, NMR lineshape analysis is one of the few techniques which can be used to study fast rate processes at equilibrium without perturbing the chemistry of the system in any way. Disturbing the nuclear spin levels involves only minute energy changes which have no effect on the chemistry. Most other measurements of fast reaction rates involve perturbing a chemical reaction from equilibrium and then watching it decay back.[23]

The study of rate processes as a function of concentrations of constituents of the system under investigation leads to rate laws and kinetics. The NMR lineshape method provides $1/\tau$, the mean lifetime of a species between successive exchanges. This quantity is related to R_{ex} the rate law[24]

$$1/\tau_{sp} = R_{ex}/(\mathrm{sp}), \tag{1-6}$$

where sp means species. Kinetic studies of an exchange process under different conditions give information about the mechanism of exchange. Mechanisms at equilibrium are of enormous interest. Ordinarily, mechanistic information comes from reactions which are far from equilibrium.

The theory of NMR lineshapes in chemically reorganizing systems has been handled at several levels of approximation, from modified Bloch equations,[8] the Kubo–Anderson–Sack treatment,[25] to the full fledged density matrix theory.[26–28]

The coupled density matrix equations, to be derived and discussed in this book, which need to be solved to obtain the absorption, have the form[28]

$$0 = \dot{\rho} = i\hbar[\rho, H] - R\rho + [\rho(\mathrm{col}) - \rho]/\tau, \tag{1-7}$$

where R is the relaxation operator. The term in brackets, which has the effect of mixing transitions and averaging resonances, is the contribution due to exchange. As will be shown later, the derivation of ρ(col) is based on the *physical description* of the *exchange mechanism* as it effects the nuclear spins. To this extent the mechanism of the exchange process is incorporated into the equations needed to calculate the NMR lineshape. So, in principle, each reaction mechanism at equilibrium leaves behind its distinctive signature. Thus, it should be possible to compare the experimental spectra with lineshapes calculated appropriate to different proposed mechanisms. Naturally, this is not to imply that every exchange process gives rise to a unique NMR lineshape!

Recently, NMR lineshape analysis has also been applied to data obtained for exchanging systems under conditions of high rf power[29, 30] and double resonance.[31–33]

2. The Bloch Equations

The first theoretical treatment of NMR considered the resonance of a large number of equivalent spins

$$\dot{\mathbf{M}}_{(x,y)} = -\gamma(\mathbf{M} \times \mathbf{H})_{(x,y)} - \mathbf{M}_{(x,y)}/T_2, \tag{1-8}$$

$$\dot{M}_z = -\gamma(\mathbf{M} \times \mathbf{H})_z - (M_z - M_z^0)/T_1 \tag{1-9}$$

where $\mathbf{H}$ represents all applied magnetic fields, $\mathbf{M}$ is the induced magnetization, T_2 and T_1 are the decay constants, respectively, of the transverse (x,y) and the longitudinal (z) components of magnetization, and γ is the gyromagnetic ratio which can be positive or negative.

The magnetic field, as we will use it, is made up of one large static component in the z direction called H_0 and a much smaller circularly rotating field of frequency ω_1 in the (x, y) plane; thus

$$\mathbf{H} = \mathbf{k}H_0 + H_1(\mathbf{i} \cos \omega t + \mathbf{j} \sin \omega t). \tag{1-10}$$

Substituting Eq. (1-10) into (1-8) and (1-9) and collecting all terms in **i**, **j**, and **k**, one obtains under steady state conditions

$$\dot{M}_x = -\omega_0 M_y + \omega_1 M_z \sin \omega t - M_x/T_2, \tag{1-11}$$

$$\dot{M}_y = -\omega_1 M_z \cos \omega t + \omega_0 M_x - M_y/T_2, \tag{1-12}$$

$$\dot{M}_z = 0 = -\omega_1 M_x \sin \omega t + \omega_1 M_y \cos \omega t - (M_z - M_z^0)/T_1, \tag{1-13}$$

where

$$\omega_0 = \gamma H_0, \qquad \omega_1 = \gamma H_1. \tag{1-14}$$

The magnetization **M** which results from solving (1-11)–(1-13) is represented in Fig. 1-1, where one sees that

$$\begin{aligned} M_x &= |m| \cos(\omega t + \varphi), \\ M_y &= |m| \sin(\omega t + \varphi), \\ M_z &= \text{independent of time,} \end{aligned} \tag{1-15}$$

and φ is the lag or phase angle between **m** and the rf field. Given the definition

$$M = M_x + iM_y = |m|e^{i(\omega t + \varphi)}, \tag{1-16}$$

$\dot{M}$ is found to be

$$\dot{M} = i\omega|m|e^{i(\omega t + \varphi)}. \tag{1-16a}$$

Adding $\dot{M}_x + i\dot{M}_y$ from (1-11) and (1-12) and using (1-16a) eliminates the

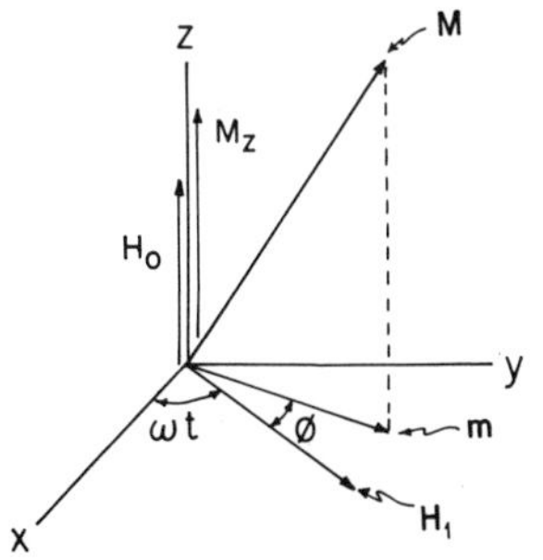

Fig. 1-1. Magnetic vectors and applied static time-dependent field under steady state conditions. $\mathbf{M} = \mathbf{m} + \mathbf{k}M_z$.

time dependence and gives

$$\left[i(\omega - \omega_0) + T_2^{-1}\right]|m| = -i\omega_1 e^{-i\varphi} M_z. \tag{1-17}$$

Next, writing out the real and imaginary part of Eq. (1-17) one has

$$(\omega - \omega_0)|m| = -\omega_1 \cos \varphi M_z, \tag{1-18a}$$

$$T_2^{-1}|m| = -\omega_1 \sin \varphi M_z. \tag{1-18b}$$

Equations (1-18a) and (1-18b) are then solved for $|m|$, by first eliminating φ to yield

$$|m| = \frac{\omega_1 M_z}{\left[(\omega - \omega_0)^2 + T_2^{-2}\right]^{1/2}}, \tag{1-19}$$

which substituted back into (1-18b) yields

$$\sin \varphi = -\frac{T_2^{-1}}{\left[(\omega - \omega_0)^2 + T_2^{-2}\right]^{1/2}}. \tag{1-20}$$

Finally, Eqs. (1-15), (1-19), and (1-20) are substituted into (1-13)† and one obtains

$$M_z = \frac{M_z^0\left[(\omega - \omega_0)^2 + T_2^{-2}\right]}{(\omega - \omega_0)^2 + T_2^{-2} + \omega_1^2(T_1/T_2)}. \tag{1-21}$$

The absorption is the component of magnetization *out of phase* with the radio frequency field. Thus, expanding M_x in (1-15) one has

$$M_x = |m|\{\cos(\omega t) \cos \varphi - \sin(\omega t) \sin \varphi\} \tag{1-22}$$

and so the absorption will be proportional to

$$\text{Abs} \sim -|m| \sin \varphi. \tag{1-23}$$

Using Eqs. (1-19)–(1-21), (1-23) becomes

$$\text{Abs} = \frac{T_2^{-1}\omega_1 M_z^0}{(\omega - \omega_0)^2 + T_2^{-2} + \omega_1^2(T_1/T_2)}, \tag{1-24}$$

giving rise to an effective linewidth of

$$T_{\text{eff}}^{-1} = \left[T_2^{-2} + \omega_1^2(T_1/T_2)\right]^{1/2}. \tag{1-25}$$

The definition of *low rf field* (low power) will be fields for which

$$\omega_1^2 T_1/T_2 \ll T_2^{-2}, \tag{1-26}$$

so at low power the effective linewidth is T_2^{-1}. One also notices in (1-21) that in this limit

$$M_z = M_z^0, \tag{1-26a}$$

†We use the relationship $\sin(x - y) = \sin x \cos y - \cos x \sin y$.

TABLE 1-1

n	0	0.1	0.2	0.3	0.4	0.5	0.6	0.7	0.8
y	1	0.990	0.960	0.911	0.843	0.762	0.671	0.578	0.488

i.e., at low power the *diagonal magnetization is given by its equilibrium value*. Thus, one sees that the transverse relaxation time determines the full width in the low power limit and that the calculation of the absorption can be simplified in the low power limit by substituting M_z^0 for M_z in the original defining equations. This simplification will be used in Chapter V.

Problems

1. Obtain the magnetization in phase with the rf field.

2. Obtain the derivative of the absorption and relate the peak separation to the linewidth.

3. Note that a single Bloch equation will always lead to a Lorentzian lineshape, $y = 1/(1 + n^2)$. Is the lineshape in Table 1-1 Lorentzian? If not, give a quantification of the difference.

REFERENCES

1. N. Bloembergen, E. M. Purcell, and R. Pound, *Phys. Rev.* **73**, 679 (1948).
2. H. S. Gutowsky and G. E. Pake, *J. Chem. Phys.* **18**, 162 (1950).
3. G. E. Pake, *J. Chem. Phys.* **16**, 327 (1948).
4. H. S. Gutowsky, G. E. Pake, and R. Bersohn, *J. Chem. Phys.* **22**, 643 (1948).
5. U. Liddel and N. F. Ramsey, *J. Chem. Phys.* **19**, 1608 (1951).
6. J. T. Arnold and M. E. Packard, *J. Chem. Phys.* **19**, 1608 (1951).
7. H. S. Gutowsky and H. Saika, *J. Chem. Phys.* **21**, 1688 (1953).
8. H. S. Gutowsky and C. H. Holm, *J. Chem. Phys.* **25**, 1288 (1957).
9. J. D. Roberts, "Nuclear Magnetic Resonance, Applications to Organic Chemistry," p. 63. McGraw Hill, New York, 1959.
10. E. Grunwald and C. F. Jumper, *J. Amer. Chem. Soc.* **85**, 2051 (1963); E. Grunwald and S. Meiboom, *J. Am. Chem. Soc.* **85**, 2047 (1963).
11. E. L. Holm and D. E. Maxwell, *Phys. Rev.* **88**, 1070 (1952).
12. I. Solomon and W. Bloembergen, *J. Chem. Phys.* **25**, 261 (1956).
13. J. T. Arnold, *Phys. Rev.* **102**, 136 (1957).
14. R. A. Ogg, *Trans. Faraday Soc.* **19**, 239 (1955).
15. L. W. Reeves, *Advan. Phys. Org. Chem.* 187 (1967); C. S. Johnson, *Adv. Magn. Reson.* **1**, 33 (1965).
16. H. Kessler, *Angew. Chem., Int. Ed. Engl.* **9**, 219 (1970).
17. F. H. Westheimer, *Acc. Chem. Res.* **1**, 90 (1968).
18. F. A. L. Anet and R. Anet, *in* "Determination of Organic Structures by Physical Methods" (F. C. Nachod and J. J. Zuckerman, eds.), Vol. 3, p. 343. Academic Press,

New York, 1971; G. L. Martin and M. L. Martin, *Prog. Nucl. Magn. Reson. Spectrosc.* **8**, 163 (1972).
19. A. T. Bottini and J. D. Roberts, *J. Amer. Chem. Soc.* **80**, 5203 (1958).
20. E. Grunwald, A. Loewenstein, and S. Meiboom, *J. Chem. Phys.* **27**, 630, 646 (1957).
21. J. P. Oliver, *Advan. Organomet. Chem.* **8**, 167 (1970).
22. R. H. Holm and M. J. O'Connor, *Progr. Inorg. Chem.* **14**, 241 (1971); N. Serpone and D. G. Bickley, *Progr. Inorg. Chem.* **17**, 391 (1972); J. J. Fortman and R. E. Sievers, *Coord. Chem. Rev.* **6**, 331 (1971).
23. E. F. Caldin, "Fast Reactions in Solution," p. 59. Blackwells, Oxford, 1964.
24. E. Grunwald, A. Loewenstein, and S. Meiboom, *J. Chem. Phys.* **27**, 630 (1957).
25. R. Kubo, *J. Phys. Soc. Japan* **9**, 935 (1954); P. W. Anderson, *J. Phys. Soc. Japan.* **9**, 316 (1954); R. H. Sack, *Mol. Phys.* **1**, 163 (1958); H. M. McConnell and C. H. Holm, *J. Chem. Phys.* **28**, 430 (1958).
26. J. I. Kaplan, *J. Chem. Phys.* **28**, 278, (1958); **28**, 462 (1958).
27. S. Alexander, *J. Chem. Phys.* **37**, 967, (1954); **38**, 1787 (1963); **40**, 2741 (1964).
28. J. I. Kaplan and G. Fraenkel, *J. Amer. Chem. Soc.* **94**, 2907 (1972).
29. R. K. Harris and W. C. Pyper, *Mol. Phys.* **21**, 467 (1971); R. K. Harris and K. M. Worvill, *J. Magn. Reson.* **9**, 294, 383 (1973).
30. J. Kaplan, P. P. Yang, and G. Fraenkel, *J. Chem. Phys.* **60**, 4840 (1974).
31. P. P. Yang and S. L. Gordon, *J. Chem. Phys.* **54**, 1779 (1971); B. M. Fung and P. M. Olympia, *Mol. Phys.* **19**, 685 (1970).
32. J. M. Anderson, *J. Magn. Reson.* **4**, 184 (1971).
33. J. I. Kaplan, P. P. Yang, and G. Fraenkel, *J. Amer. Chem. Soc.* **97**, 3881 (1975).

GENERAL REFERENCES

F. Bloch, *Phys. Rev.* **70**, 460 (1946).
F. Bloch, W. W. Hansen, and M. Packard, *Phys. Rev.* **70**, 444 (1946).
E. M. Purcell, H. C. Torrey, and R. N. Pound, *Phys. Rev.* **69**, 37 (1946).
A. K. Saha and T. D. Das, "Theory and Applications of Nuclear Induction," Saha Inst. Nucl. Phys., Calcutta, 1957.

Chapter II
DENSITY MATRIX

1. The Density Matrix Equation

In describing nuclear magnetic resonance we must simultaneously bring in the quantum mechanical aspects of the problem (the Hamiltonian, energy levels, matrix elements, *etc.*) and the statistical aspects (temperature, distribution probability).

First, we characterize our system with the Hamiltonian $\mathcal{H}_A$. $\mathcal{H}_A$ is thought to encompass those parts of the system which are interacting with one another. Thus, for a very dilute pure gas $\mathcal{H}_A$ would consist only of the Hamiltonian of a single molecule. As the gas density goes up $\mathcal{H}_A$ would contain the molecular Hamiltonian plus terms describing the interactions between the molecules. If the gas collided with the walls of the container $\mathcal{H}_A$ would have to include the container Hamiltonian as well as the gas molecule–wall interaction.

At a given time t, we imagine a very large number N of duplicate systems, A_i, with wave functions $\psi_i^A(t)$. Each such $\psi_i^A(t)$ obeys the time-dependent Schrödinger equation

$$i\hbar\dot{\psi}_i^A(t) = \mathcal{H}_A\psi_i(t) \tag{2-1}$$

with the normalization

$$\int \psi_i^{A*}(t)\psi_i^A(t)\,d\tau = 1. \tag{2-2}$$

Equation (2-1) can also be written in Dirac notation as

$$i\hbar\frac{d}{dt}|\psi_i^A(t)\rangle = \mathcal{H}_A|\psi_i^A(t)\rangle. \tag{2-3}$$

A matrix element of an arbitrary operator A taken between states ϕ_1 and ϕ_2 can be written as (* means complex conjugate)

$$A_{1,2} = \int \phi_1^* \mathbf{A} \phi_2 \, d\tau = \langle \phi_1|\mathbf{A}|\phi_2\rangle. \tag{2-4}$$

The density matrix is defined as the operator

$$\rho = \frac{1}{N}\sum_{i=1}^{N} |\psi_i^A(t)\rangle\langle\psi_i^A(t)|, \tag{2-5}$$

and thus the matrix element between the states ϕ_1 and ϕ_2 is given as

$$\rho_{1,2} = \int \phi_1^* \rho \phi_2 \, d\tau = \langle\phi_1|\rho|\phi_2\rangle \tag{2-6}$$

$$= \frac{1}{N}\sum_{i=1}^{N} \langle\phi_1|\psi_i^A(t)\rangle\langle\psi_i^A(t)|\phi_2\rangle. \tag{2-7}$$

Next, we express $\psi_i^A(t)$ as an orthogonal set of states ϕ_n of which ϕ_1 and ϕ_2 are members; thus

$$\psi_i^A(t) = \sum_n a_n^i(t)\phi_n. \tag{2-8}$$

The requirement of an orthogonal set is that

$$\int \phi_n^* \phi_m \, d\tau = \delta_{n,m},$$

where

$$\delta_{n,m} = \begin{cases} 1, & n = m \quad (2\text{-}9) \\ 0, & n \neq m. \quad (2\text{-}10) \end{cases}$$

Substituting (2-8) into (2-7) and using (2-9) gives us

$$\rho_{1,2} = \langle\phi_1|\rho|\phi_2\rangle = \frac{1}{N}\sum_{i=1}^{N} a_1^i a_2^{i*}. \tag{2-11}$$

Note,

$$\rho_{1,2} = \rho_{2,1}^*,$$

which means that ρ is hermitian. In (2-11) we can see the combined quantum mechanical (the projection $a_1^i a_2^{i*}$) and the statistical (weighted sum over i) nature of the density matrix.

From (2-11) we can obtain the result that

$$\operatorname{Tr}\rho = \sum_n \langle\phi_n|\rho|\phi_n\rangle = 1. \tag{2-12}$$

Proof:

$$\operatorname{Tr}\rho = \frac{1}{N}\sum_i^N \sum_n a_n^i a_n^{i*}. \tag{2-13}$$

The normalization of ψ_i^A [(2-9)] requires that

$$\int \psi_i^{A*}(t)\psi_i^A(t)\,d\tau = \sum_{n,m} a_n^{i*}a_m^i \int \phi_n^*\phi_m\,d\tau \tag{2-14}$$

$$= \sum_n a_n^{i*}a_n^i = 1, \tag{2-15}$$

where we have used (2-10). Substituting (2-15) into (2-13) gives

$$\operatorname{Tr}\rho = \frac{1}{N}\sum_{i=1}^N 1(i) = 1. \tag{2-16}$$

So far we have defined ρ [(2-5)] and described its form in an orthogonal representation [(2-11)]. Next, we must obtain a differential equation for ρ. To do this we substitute $\psi_i^A(t)$, as given in (2-8), into (2-1). Then, using the orthogonal relationship, given in (2-9), one obtains

$$i\hbar\dot{a}_n^i = \sum_m \langle\phi_n|\mathcal{H}^A|\phi_m\rangle a_m^i. \tag{2-17}$$

Next, we take the time derivative of Eq. (2-11) and obtain

$$\dot{\rho}_{1,2} = \frac{1}{N}\sum_i \left[\dot{a}_1^i a_2^{i*} + a_1^i \dot{a}_2^{i*}\right]. \tag{2-18}$$

Applying the complex conjugate operation to (2-17) one obtains

$$-i\hbar\dot{a}_n^{i*} = \sum_m \langle\phi_n|\mathcal{H}^A|\phi_m\rangle^* a_m^{i*}. \tag{2-19}$$

Note $\mathcal{H}^A$ is a hermitian operator with real eigenvalues and has the property that

$$\langle\phi_n|\mathcal{H}^A|\phi_m\rangle = \langle\phi_m|\mathcal{H}^A|\phi_n\rangle^*. \tag{2-20}$$

Substituting (2-20) into (2-19) then gives us

$$-i\hbar\dot{a}_n^{i*} = \sum_m \langle\phi_m|\mathcal{H}^A|\phi_n\rangle a_m^{i*}. \tag{2-21}$$

Then, substituting into (2-18) for a_n^i and a_n^{i*}, as given in (2-17) and (2-20), one has

$$\dot{\rho}_{1,2} = -\frac{i\hbar^{-1}}{N} \sum_i \sum_m \left[\mathcal{H}^A_{1,m} a^i_m a^{i*}_2 - a^i_1 \mathcal{H}^A_{m,2} a^{i\,*}_m \right], \tag{2-22}$$

$$\dot{\rho}_{1,2} = -i\hbar^{-1} \sum_m \left[\mathcal{H}^A_{1,m} \rho_{m,2} - \rho_{1,m} \mathcal{H}^A_{m,2} \right], \tag{2-23}$$

or

$$\langle \phi_1 | \{ \dot{\rho} = -i\hbar [\mathcal{H}^A, \rho] \} | \phi_2 \rangle, \tag{2-24}$$

where we have used Eq. (2-11) on going from (2-22) to (2-23) and also used the product operator relation

$$\langle 1 | \mathbf{AB} | 2 \rangle = \sum_n \langle 1 | \mathbf{A} | n \rangle \langle n | \mathbf{B} | 2 \rangle. \tag{2-25}$$

Finally, since ϕ_1 and ϕ_2 are arbitrary we have the general operator relation

$$\dot{\rho} = -i\hbar^{-1} [\mathcal{H}^A, \rho], \tag{2-26}$$

where $[\mathbf{A}, \mathbf{B}] = \mathbf{AB} - \mathbf{BA}$. Equation (2-26) is the principal result of this chapter.

2. Average Values of Operators

Given ρ how do we calculate average values of operators? To derive this we start with an ensemble average value of A given as

$$\langle A \rangle = \frac{1}{N} \sum_i \langle \psi_i^A | A | \psi_i^A \rangle. \tag{2-27}$$

Substituting for ψ_i^A in (2-27) using (2-8) one obtains

$$\langle A \rangle = \frac{1}{N} \sum_{i=1}^{N} a^i_m a^{i*}_n \langle \phi_n | A | \phi_m \rangle \tag{2-28}$$

$$= \operatorname{Tr} \rho A, \tag{2-29}$$

where going from (2-28) to (2-29) we have used (2-11) and (2-25).

To see how Eq. (2-26) is used, imagine that $\mathcal{H}^A$ represents the Hamiltonian for a single spin I acted on by the magnetic field

$$\mathbf{H} = \mathbf{k} H_0 + \mathbf{i} H_1 \cos \omega t + \mathbf{j} H_1 \sin \omega t. \tag{2-30}$$

The time-dependent part of H is circularly polarized (we will use this form of the rf field quite generally in all of this book) but any time-dependent field could be used in the following discussion. Then, $\mathcal{H}^A$ will be given as

$$\hbar^{-1} \mathcal{H}^A = \omega_0 I^z + \omega_1 (I^x \cos \omega t + I^y \sin \omega t). \tag{2-31}$$

The classic equation of motion for the magnetization is

$$\dot{\mathbf{M}} = -\gamma(\mathbf{M} \times \mathbf{H}). \tag{2-32}$$

We want to show that the density matrix equation for $\mathcal{H}^A$, given in (2-26), leads to exactly the same result. Let us calculate $\dot{M}_x$. From (2-32)

$$\dot{M}_x = -\gamma[M_y H_z - M_z H_y] = -\gamma[M_y H_0 - H_1 M_z \sin \omega t]. \tag{2-33}$$

From the density matrix equation (2-26) and the definition of an average value [(2-29)] we have

$$\dot{M}_x = \mathrm{Tr}\, I^x \dot{\rho} = -i\hbar^{-1}\mathrm{Tr}\, I^x[\mathcal{H}^A, \rho]. \tag{2-34}$$

We now make use of the result that (see Appendix to this chapter)

$$\mathrm{Tr}\, ABC = \mathrm{Tr}\, CAB = \mathrm{Tr}\, BCA. \tag{2-35}$$

Therefore, using (2-35) we can rewrite (2-34) as

$$\dot{M}_x = -i\hbar^{-1}\mathrm{Tr}\, \rho[I^x \mathcal{H}^A - \mathcal{H}^A I^x] = -i\hbar^{-1}\mathrm{Tr}\, \rho[I^x, \mathcal{H}^A]. \tag{2-36}$$

Using the commutation rules of angular momentum

$$[I^x, I^y] = i\hbar I^z, \tag{2-37}$$

$$[I^x, I^z] = -i\hbar I^y, \tag{2-38}$$

and (2-31) to define $\mathcal{H}^A$ one then has

$$\dot{M}_x = -i\, \mathrm{Tr}\, \rho\{-i\omega_0 I^y + i\not{\omega}_1 I^z \sin \omega t\} = -\omega_0 M_y + \not{\omega}_1 M_z \sin \omega t. \tag{2-39}$$

Equation (2-39) is identical to (2-33) after identifying

$$\gamma H_0 = \omega_0, \qquad \gamma H_1 = \not{\omega}_1. \tag{2-40}$$

For practice it is suggested that the reader evaluate and compare $\dot{M}_y$ and $\dot{M}_z$ as obtained from the classical and matrix routes.

Note, in addition, that if we had modified (2-26) phenomenologically to include the interaction of the spin with its environment by writing

$$\dot{\rho} = -i\hbar^{-1}[\mathcal{H}^A, \rho] - (\rho - \rho_0)/T_1 \tag{2-41}$$

and proceeded as before, from (2-34), we would have constructed the Bloch equations (as used in Chapter I) in the form

$$\dot{\mathbf{M}} = -\gamma(\mathbf{M} \times \mathbf{H}) - (\mathbf{M} - \mathbf{M}_0)/T_1. \tag{2-42}$$

3. Summary

We have derived the density matrix equation

$$\dot{\rho} = -i\hbar^{-1}[\mathcal{H}^A, \rho]. \tag{2-26}$$

The average value of an operator A is given as

$$\langle A \rangle = \mathrm{Tr}\, \rho A. \tag{2-29}$$

Appendix

To prove

$$\mathrm{Tr}\, AB = \mathrm{Tr}\, BA. \tag{2-A1}$$

Proof:

$$\mathrm{Tr}\, AB = \sum_{n,m} \langle n|\mathbf{A}|m\rangle\langle m|B|n\rangle$$

$$= \sum_{n,m} \langle m|B|n\rangle\langle n|A|m\rangle$$

$$= \mathrm{Tr}\, BA. \tag{2-A2}$$

To prove

$$\mathrm{Tr}\, ABC = \mathrm{Tr}\, CAB, \tag{2-35}$$

let

$$G = AB. \tag{2-A3}$$

From (2-A1) we have

$$\mathrm{Tr}\, GC = \mathrm{Tr}\, CG \tag{2-A4}$$

and thus

$$\mathrm{Tr}\, ABC = \mathrm{Tr}\, CAB. \tag{2-35}$$

Problems

1. Solve for $\dot{M}_y$ and $\dot{M}_z$ using the classical and density matrix procedures.

2. Solve for $M_z(t)$ when it is set equal to $-M_z^0$ at $t = 0$. This is the so-called "inversion-recovery" method for obtaining T_1.

GENERAL REFERENCES

P. A. M. Dirac, "Principles of Quantum Mechanics," 3rd Ed., p. 222. Oxford Univ. Press, London, 1947.

U. Fano, *Rev. Mod. Phys.* **29**, 74 (1957).

J. Liouville, *J. Math.* **3**, 349 (1838).

L. Landau, *Z. Phys.* **45**, 430 (1927).

R. McWeeny, *Rev. Mod. Phys.* **32**, 335 (1960).

C. P. Slichter, "Principles of Magnetic Resonance," Chap. 5. Harper, New York, 1964.

R. Tolman, "Statistical Mechanics," Chap. 9. Oxford Univ. Press, London, 1930.

J. von Neumann, *Gottinger Nachr.* p. 246 (1924).

Chapter III

DENSITY MATRIX TREATMENT OF THE SINGLE ABSORPTION PEAK

In Chapter I we derived the expression for the NMR absorption due to an assembly of noninteracting spins. Now, we will repeat the calculation using the density matrix approach, the purpose being to familiarize the reader with the notation and procedures to be used throughout this book. We start then with the density matrix equation appropriate for noninteracting spins acted on by the magnetic field[†] [(1-10)]

$$\dot{\rho} = -i\left[\omega_0 I^z + \omega_1 I^x \cos \omega t + \omega_1 I^y \sin \omega t, \rho\right] - T^{-1}(\rho - \rho_0), \quad (3\text{-}1)$$

where

$$\rho_0 = e^{-\hbar\omega_0 I^z/kT_b} / \mathrm{Tr}\, e^{-\hbar\omega_0 I^z/kT_b}. \quad (3\text{-}2)$$

Our relaxation form is equivalent to setting $T_1 = T_2$ in (1-8) and (1-9). This will make the calculation simpler than the more general form which will be introduced in Chapter IV.

The representation we will use is one for which I^z is diagonalized, i.e.,

$$I^z|m\rangle = m|m\rangle. \quad (3\text{-}3)$$

For spin $\frac{1}{2}$ the states are commonly written as

$$|\tfrac{1}{2}\rangle = \alpha = \begin{bmatrix} 1 \\ 0 \end{bmatrix}, \qquad |-\tfrac{1}{2}\rangle = \beta = \begin{bmatrix} 0 \\ 1 \end{bmatrix} \quad (3\text{-}4)$$

[†]Notice the constant in front of the bracket is $-i$ since ω is in rad/sec; $-i\hbar^{-1}[\mathcal{H}, \rho]$ applies when $\mathcal{H}$ has the dimension of energy.

with the matrix representation in these states for the spin operators given as

$$I^z = \frac{1}{2}\begin{bmatrix} 1 & 0 \\ 0 & -1 \end{bmatrix}, \quad I^x = \frac{1}{2}\begin{bmatrix} 0 & 1 \\ 1 & 0 \end{bmatrix}, \quad I^y = \frac{1}{2}\begin{bmatrix} 0 & -i \\ i & 0 \end{bmatrix}. \tag{3-5}$$

It is simpler to work with the raising and lowering operators

$$\begin{aligned} I^+ &= I^x + iI^y \\ I^- &= I^x - iI^y. \end{aligned} \tag{3-6}$$

For spin $\frac{1}{2}$ in the α, β representation these operators are

$$I^+ = \begin{bmatrix} 0 & 1 \\ 0 & 0 \end{bmatrix}, \qquad I^- = \begin{bmatrix} 0 & 0 \\ 1 & 0 \end{bmatrix} \tag{3-7}$$

and they act on α and β as

$$\begin{aligned} I^+|\alpha &= \begin{bmatrix} 0 & 1 \\ 0 & 0 \end{bmatrix}\begin{bmatrix} 1 \\ 0 \end{bmatrix} = \begin{bmatrix} 0 \\ 0 \end{bmatrix} = 0, \\ I^-|\alpha &= \begin{bmatrix} 0 & 0 \\ 1 & 0 \end{bmatrix}\begin{bmatrix} 1 \\ 0 \end{bmatrix} = \begin{bmatrix} 0 \\ 1 \end{bmatrix} = \beta, \\ I^+|\beta &= \begin{bmatrix} 0 & 1 \\ 0 & 0 \end{bmatrix}\begin{bmatrix} 0 \\ 1 \end{bmatrix} = \begin{bmatrix} 1 \\ 0 \end{bmatrix} = \alpha, \\ I^-|\beta &= \begin{bmatrix} 0 & 0 \\ 1 & 0 \end{bmatrix}\begin{bmatrix} 0 \\ 1 \end{bmatrix} = \begin{bmatrix} 0 \\ 0 \end{bmatrix}. \end{aligned} \tag{3-8}$$

Rewriting (3-2) in the I^+, I^- representation one obtains (true for any I)

$$\dot{\rho} = -i\left[\omega_0 I^z + \frac{\omega_1}{2} I^+ e^{-i\omega t} + \frac{\omega_1}{2} I^- e^{i\omega t}, \rho\right] - \frac{\rho - \rho_0}{T}. \tag{3-9}$$

For normal temperatures (i.e., $T > 1°\mathrm{K}$)

$$\hbar\omega_0/kT \ll 1, \tag{3-10}$$

therefore one can accurately approximate ρ_0 as

$$\rho_0 \sim \frac{\mathcal{I} - \hbar\omega_0 I^z/kT_b}{2I + 1}. \tag{3-11}$$

Next, we simplify (3-9) by eliminating the time dependence by going into the rotating coordinate system. In the laboratory frame the rf field can be thought of as a radial line on a rotating phonograph record. Going into the rotating frame is equivalent to hopping on the record. The transformed density matrix operator $\tilde{\rho}(t)$ in the rotating coordinate system is given as[1]

$$\tilde{\rho}(t) = \mathcal{U}\rho\mathcal{U}^{-1}, \tag{3-12}$$

where

$$\mathcal{U} = e^{i\omega I^z t}. \tag{3-13}$$

Thus, we obtain

$$\dot{\rho} = \frac{d}{dt}\{e^{i\omega I^z t}\tilde{\rho} e^{i\omega I^z t}\}$$
$$= \left[-i\omega I^z, e^{-i\omega I^z t}\tilde{\rho} e^{i\omega I^z t}\right] + e^{-i\omega I^z t}\dot{\tilde{\rho}} e^{i\omega I^z t}. \tag{3-14}$$

Substituting (3-14) into (3-9) gives the result

$$\dot{\tilde{\rho}} = -i\left[(\omega_0 - \omega)I^z + \frac{\omega_1}{2} e^{i\omega I^z t} I^+ e^{-i\omega I^z t} e^{-i\omega t} + \frac{\omega_1}{2} e^{i\omega I^z t} I^- e^{-i\omega I^z t} e^{i\omega t}, \tilde{\rho}\right] + \frac{\tilde{\rho} - e^{i\omega I^z t}\rho_0 e^{-i\omega I^z t}}{T}. \tag{3-15}$$

First, note that

$$e^{i\omega I^z t}\rho_0 e^{-i\omega I^z t} = \rho_0 \tag{3-16}$$

as any operator (I^z) commutes with any function (ρ_0) of this operator. Second, we must evaluate

$$e^{i\omega I^z t} I^{\pm} e^{-i\omega I^z t} = G_{\pm}. \tag{3-17}$$

We do this by forming the differential equation

$$\dot{G}_{\pm} = i\omega e^{i\omega I^z t}[I^z, I^{\pm}]e^{-i\omega I^z t}. \tag{3-18}$$

Now, as

$$[I^z, I^{\pm}] = \pm I^{\pm}, \tag{3-19}$$

it follows from (3-18) that

$$\dot{G}_{\pm} = \pm i\omega G_{\pm}. \tag{3-20}$$

Integrating (3-20) with the initial condition at $t = 0$ [see (3-17)] that

$$G_{\pm}(0) = I^{\pm}, \tag{3-21}$$

one obtains

$$G_{\pm} = I^{\pm} e^{\pm i\omega t}. \tag{3-22}$$

Substituting (3-22) into (3-15) one then has

$$\dot{\tilde{\rho}} = -i\left[(\omega_0 - \omega)I^z + \frac{\omega_1}{2}(I^+ + I^-), \tilde{\rho}\right] - \frac{\tilde{\rho} - \rho_0}{T}. \tag{3-23}$$

As we are looking for the steady state solution and since the right-hand side is independent of time, it follows that $\dot{\tilde{\rho}} = 0$.

There are two approaches we can take to solve (3-23). We can either represent $\tilde{\rho}$ as

$$\tilde{\rho} = \rho_0 + \rho_1 \tag{3-24}$$

(the more customary usage) or as

$$\tilde{\rho} = N^{-1} + \rho_{1'}, \tag{3-25}$$

where N is the number of states $(2I + I)$ [see (3-11)]. We have used the form (3-25) at high power and for double resonance because it avoided some computer cancellation difficulties we encountered with (3-24). Formally one can employ (3-24) for all purposes. For the low power solutions (3-24) is the appropriate equation. It expresses the solution at low power in a form which clearly shows that the perturbed diagonal matrix elements of Eq. (3-23) are not needed. This follows because at low power the diagonal elements of $\tilde{\rho}$ are just the diagonal elements of ρ_0. Thus, the $\rho_0 + \rho_1$ form will be used for low power in Chapter VI, while $N^{-1} + \rho_{1'}$ is used for high power conditions in Chapters VII and VIII.

To obtain the absorption due to an assembly of single spins, $I = \frac{1}{2}$, we choose here the form of $\tilde{\rho} = N^{-1} + \rho_{1'}$. Substituting (3-25) into (3-23) we obtain the equation

$$\dot{\tilde{\rho}} = 0 = -i\left[(\omega_0 - \omega)I^z + \frac{\not{\omega}_1}{2}(I^+ + I^-), \rho_{1'}\right] - \frac{\rho_{1'} + bI^z}{T}, \tag{3-26}$$

using (3-11) for ρ_0, where

$$b = \hbar\omega_0 / NkT_b \tag{3-27}$$

and T_b is the temperature.

Next, we evaluate Eq. (3-26) in the α, β representation. Calling α and β, 1 and 2, respectively, we find the $\langle 1 \| 2 \rangle$ element of (3-26) to be

$$-i(\omega_0 - \omega)\langle 1|\rho_{1'}|2\rangle - \frac{i\not{\omega}_1}{2}[\langle 2|\rho_{1'}|2\rangle - \langle 1|\rho_{1'}|1\rangle] - \frac{\langle 1|\rho_{1'}|2\rangle}{T} = 0. \tag{3-28}$$

As $\tilde{\rho}$ and thus $\rho_{1'}$ are hermitian operators, i.e.,

$$\langle 2|\rho_{1'}|1\rangle = \langle 1|\rho_{1'}|2\rangle^*, \tag{3-29}$$

this enables us to take the complex conjugate of (3-28) as

$$i(\omega_0 - \omega) < 2|\rho_{1'}|1\rangle + \frac{i\not{\omega}_1}{2}[\langle 2|\rho_{1'}|2\rangle - \langle 1|\rho_{1'}|1\rangle] - \frac{\langle 2|\rho_{1'}|1\rangle}{T} = 0. \tag{3-30}$$

The other equations needed are the $\dot{\tilde{\rho}}_{1,1}$ and $\dot{\tilde{\rho}}_{2,2}$ density matrix elements, respectively, giving the equations

$$0 = -\frac{i\not{\omega}_1}{2}[\langle 2|\rho_{1'}|1\rangle - \langle 1|\rho_{1'}|2\rangle] - \frac{\langle 1|\rho_{1'}|1\rangle}{T} - \frac{b}{2T} \tag{3-31}$$

$$0 = -\frac{i\not{\omega}_1}{2}[\langle 1|\rho_{1'}|2\rangle - \langle 2|\rho_{1'}|1\rangle] - \frac{\langle 2|\rho_{1'}|2\rangle}{T} + \frac{b}{2T}. \tag{3-32}$$

Letting

$$\omega - \omega_0 = \Delta\omega$$

we can write Eqs. (3-28), (3-30), (3-31), and (3-32) in matrix form as

$$\begin{bmatrix} +i\Delta\omega - 1/T & 0 & +i\omega_1/2 & -i\omega_1/2 \\ 0 & -i\Delta\omega - 1/T & -i\omega_1/2 & +i\omega_1/2 \\ +i\omega_1/2 & -i\omega_1/2 & -1/T & 0 \\ -i\omega_1/2 & +i\omega_1/2 & 0 & -1/T \end{bmatrix} \begin{bmatrix} (\rho_{1'})_{1,2} \\ (\rho_{1'})_{2,1} \\ (\rho_{1'})_{1,1} \\ (\rho_{1'})_{2,2} \end{bmatrix} = \begin{bmatrix} 0 \\ 0 \\ +b/2T \\ -b/2T \end{bmatrix} \tag{3-33}$$

We strongly suggest that the reader verify this matrix.

As stated previously the absorption is proportional to the component of magnetization out of phase with the applied rf field. In the rotating coordinate system the rf field is in the x direction, so it then follows that the absorption will be given as

$$\begin{aligned} \text{Abs} \sim \overline{M_y} &= \text{Tr}\, \tilde{\rho} I^y \\ &= \text{Tr}\, N^{-1} I^y + \text{Tr}\, \rho_{1'} I^y \\ &= -\tfrac{1}{2} i\big[\langle 2|\rho_{1'}|1\rangle - \langle 1|\rho_{1'}|2\rangle\big]. \end{aligned} \tag{3-34}$$

Now, since $\rho_{1'}$ is hermitian, the absorption becomes

$$\text{Abs} \sim \overline{M_y} = -\text{Im}\langle 1|\rho_{1'}|2\rangle. \tag{3-35}$$

In general, to obtain the $(\rho_{1'})_{1,2}$ element we must solve the set of four coupled equations given in (3-33). For just the low power absorption, i.e., for

$$\langle 1|\rho_{1'}|2\rangle \sim \omega_1,$$

we solve (3-31) and (3-32) to zero order in ω_1 as

$$\begin{aligned} \langle 1|\rho_{1'}|1\rangle &= -\tfrac{1}{2} b, \\ \langle 2|\rho_{1'}|2\rangle &= +\tfrac{1}{2} b, \end{aligned} \tag{3-36}$$

and substitute these results into (3-28) to give

$$(\rho_{1'})_{1,2} = -\tfrac{1}{2} i\omega_1 b / \big[i(\omega_0 - \omega) + T^{-1}\big], \tag{3-37}$$

and the absorption will be obtained from (3-35) as

$$\begin{aligned}
\text{Abs} \sim -\text{Im}\left[\frac{-\frac{1}{2}i\omega_1 b}{i(\omega_0-\omega)+T^{-1}}\right] \\
= \text{Im}\,\frac{1}{2}\left[\frac{i\omega_1 b(-i(\omega_0-\omega)+T^{-1})}{(\omega_0-\omega)^2+T^{-2}}\right] \\
= \frac{\omega_1 b/2T}{(\omega_0-\omega)^2+T^{-2}}.
\end{aligned} \qquad (3\text{-}38)$$

At this point, before going on to the general solution of (3-33) let us establish in our minds, at least for the single spin example, the relationship between matrix elements of $\rho_{1'}$ and the $\overline{M}_x$, $\overline{M}_y$, and $\overline{M}_z$ components of magnetization. In analogy with (3-34) we find

$$\overline{M}_x = \tfrac{1}{2}\left[(\rho_{1'})_{1,2} + (\rho_{1'})_{2,1}\right], \qquad (3\text{-}39a)$$

$$\overline{M}_y = \frac{1}{2i}\left[(\rho_{1'})_{2,1} - (\rho_{1'})_{1,2}\right], \qquad (3\text{-}39b)$$

$$\overline{M}_z = \tfrac{1}{2}\left[(\rho_{1'})_{1,1} - (\rho_{1'})_{2,2}\right]. \qquad (3\text{-}39c)$$

Thus, from (3-39) we see that

$$(\rho_{1'})_{1,2} = \overline{M}_x - i\overline{M}_y\,. \qquad (3\text{-}40)$$

Equation (3-28) can thus be rewritten

$$-i(\omega_0-\omega)\left(\overline{M}_x - i\overline{M}_y\right) + i\omega_1\overline{M}_z - T^{-1}\left(\overline{M}_x - i\overline{M}_y\right) = 0, \qquad (3\text{-}41)$$

which is just a combination of two of the Bloch equations given in Chapter I. In addition, we showed in Chapter I that for low power $\overline{M}_z = \overline{M}_z^0$. To prove this here we substitute from (3-36) into (3-39c) to give

$$\overline{M}_z = -\tfrac{1}{2}b. \qquad (3\text{-}42)$$

Compare this with the definition of $\overline{M}_z^0$ which is

$$\begin{aligned}
\overline{M}_z^0 &= \text{Tr}\, I^z\rho_0 \\
&= \text{Tr}\, I^z\left[\tfrac{1}{2} - I^z b\right] \\
&= -b\,\text{Tr}\,(I^z)^2 = -\tfrac{1}{2}b.
\end{aligned} \qquad (3\text{-}43)$$

For arbitrary rf power we can represent Eq. (3-33) as

$$A_T\tilde{\rho} = B, \qquad (3\text{-}44)$$

where A is the 4×4 coefficient matrix and $\tilde{\rho}$ and B are column vectors. We solve (3-44) as

$$\tilde{\rho} = A_{\mathrm{T}}^{-1}B. \tag{3-45}$$

Thus, it appears that for each value of $\omega_0 - \omega$ we must invert a 4×4 matrix. To calculate the absorption for a large number of points this procedure can be quite time consuming. Instead, we write (3-44)

$$\begin{bmatrix} i\,\Delta\omega\,\mathcal{I} + A_1 & F \\ C & D \end{bmatrix}\begin{bmatrix} \rho_{\mathrm{I}} \\ \rho_{\mathrm{II}} \end{bmatrix} = \begin{bmatrix} 0 \\ G \end{bmatrix}, \tag{3-46}$$

where A_1, B, C, and D are 2×2 matrices and ρ_{I}, ρ_{II}, and G are 1×2 column vectors, making an identification of terms from (3-33).

To obtain the absorption we only need

$$\rho_{\mathrm{I}} = \begin{bmatrix} (\rho_{\mathrm{I}'})_{1,2} \\ (\rho_{\mathrm{I}'})_{2,1} \end{bmatrix}. \tag{3-47}$$

With this in mind we expand (3-46) to give

$$(i\,\Delta\omega\,\mathcal{I} + A_1)\rho_{\mathrm{I}} + F\rho_{\mathrm{II}} = 0, \tag{3-48a}$$

$$C\rho_{\mathrm{I}} + D\rho_{\mathrm{II}} = G. \tag{3-48b}$$

Next, we solve (3-48b) for ρ_{II} as

$$\rho_{\mathrm{II}} = -D^{-1}C\rho_{\mathrm{I}} + D^{-1}G \tag{3-49}$$

and substitute the result into (3-48a),

$$\left[i\,\Delta\omega\,\mathcal{I} + A_1 - FD^{-1}C\right]\rho_{\mathrm{I}} = -FD^{-1}G. \tag{3-50}$$

To simplify notation we define

$$A_1 - FD^{-1}C = \mathcal{A}, \qquad -FD^{-1}G = \mathcal{G}. \tag{3-51}$$

What we have done is replace the 4×4 matrix equation for $\tilde{\rho}$ with a 2×2 matrix equation for ρ_i alone, given as

$$\left[i\,\Delta\omega\,\mathcal{I} + \mathcal{A}\right]\rho_{\mathrm{I}} = \mathcal{G}. \tag{3-52}$$

Next, we define a 2×2 unitary matrix U such that

$$U\mathcal{A}U^{-1} = \mathcal{D}, \tag{3-53}$$

where $\mathcal{D}$ is diagonal. For a moment let us put aside the question of how to obtain U. As

$$U\mathcal{I}U^{-1} = \mathcal{I}, \tag{3-54}$$

we see that on substituting (3-53) into (3-50) we have

$$U\left[i\,\Delta\omega\,\mathcal{I} + \mathcal{D} \right] U^{-1} \rho_{\mathrm{I}} = \mathcal{G} \tag{3-55}$$

or

$$\rho_{\mathrm{I}} = U\left[i\,\Delta\omega\,\mathcal{I} + \mathcal{D} \right] U^{-1} \mathcal{G}, \tag{3-56}$$

which is just the sought for result. Note that once $\mathcal{D}$ is obtained, $[i\,\Delta\omega\,\mathcal{I} + \mathcal{D}]^{-1}$ is known for each $\Delta\omega$.[2]

Now let us go back and see how U is obtained. We wish to diagonalize $\mathcal{A}$ so we consider the eigenvalue equation

$$\mathcal{A}\psi_\alpha = E_\alpha \psi_\alpha \tag{3-57}$$

Equation (3-57) in the specific orthogonal representation in which $\mathcal{A}$ [see Eq. (3-52)] is given appears as

$$\sum_j \langle i|\mathcal{A}|j\rangle\langle j|\psi_\alpha\rangle = E_\alpha \langle i|\psi_\alpha\rangle. \tag{3-58}$$

Calling $\langle j|\psi_\alpha\rangle = C_\alpha^j$ one then has

$$\begin{bmatrix} A_{11} & A_{12} & \cdots & A_{1n} \\ \vdots & & & \vdots \\ A_{n1} & & & A_{nn} \end{bmatrix} \begin{bmatrix} C_\alpha^1 \\ C_\alpha^2 \\ C_\alpha^n \end{bmatrix} = E_\alpha \begin{bmatrix} C_\alpha^1 \\ C_\alpha^2 \\ C_\alpha^n \end{bmatrix}. \tag{3-59}$$

Now repeat this process for each eigenvalue. For all eigenvalues we can thus write

$$\begin{bmatrix} A_{11} & A_{12} & \cdots & A_{1n} \\ \vdots & & & \vdots \\ & & & \\ A_{n1} & & & A_{nn} \end{bmatrix} \begin{bmatrix} C_1^1 & C_2^1 & \cdots & C_n^1 \\ C_1^2 & & & \\ \vdots & & & \vdots \\ C_1^n & & & C_n^n \end{bmatrix} = \begin{bmatrix} E_1 C_1^1 & \cdots & E_n C_n^1 \\ E_1 C_1^2 & & \\ \vdots & & \vdots \\ E_1 C_1^n & \cdots & E_n C_n^n \end{bmatrix}. \tag{3-60}$$

From the orthogonality of the wave functions,

$$\int \psi_\alpha^* \psi_{\alpha'}\, d\tau = \delta_{\alpha,\alpha'}, \tag{3-61}$$

it follows that

$$\sum_j C_\alpha^{j*} C_{\alpha'}^j = \delta_{\alpha,\alpha'}, \tag{3-62}$$

and as a consequence one makes the identification that

$$U = \begin{bmatrix} C_1^1 & \cdots & C_n^1 \\ C_1^2 & & \\ \vdots & & \vdots \\ C_1^n & \cdots & C_n^n \end{bmatrix}, \tag{3-63}$$

$$U^{-1} = \begin{bmatrix} C_1^{1*} & \cdots & C_1^{n*} \\ C_2^{1*} & & \\ \vdots & & \vdots \\ C_n^{1*} & \cdots & C_n^{n*} \end{bmatrix}, \tag{3-64}$$

where $U^{-1}U = \mathcal{I}$ and

$$U^{-1}\mathcal{A}U = \begin{bmatrix} E_1 & & 0 \\ & \ddots & \\ 0 & & E_n \end{bmatrix}. \tag{3-65}$$

Equation (3-65) can be checked by straightforward multiplication and making use of (3-60) and (3-62).

Luckily one never has to obtain U, U^{-1}, and $\mathcal{D}$ by hand. Computer programs do the job.

Going back to our 2×2 space for calculating ρ_{I} for a single spin and substituting for U and U^{-1} in (3-56) we have

$$\rho_{\mathrm{I}} = \begin{bmatrix} C_1^{1*} & C_1^{2*} \\ C_2^{1*} & C_2^{2*} \end{bmatrix} \begin{bmatrix} \dfrac{1}{i\,\Delta\omega + \mathcal{D}_{11}} & 0 \\ 0 & \dfrac{1}{i\,\Delta\omega + \mathcal{D}_{22}} \end{bmatrix} \begin{bmatrix} C_1^1 & C_2^1 \\ C_1^2 & C_2^2 \end{bmatrix} \begin{bmatrix} \mathcal{G}_1 \\ \mathcal{G}_2 \end{bmatrix} \tag{3-66}$$

$$\rho_{\mathrm{I}} = \begin{bmatrix} \mathcal{G}_1\left[\dfrac{C_1^{1*}C_2^1}{i\,\Delta\omega + \mathcal{D}_{11}} + \dfrac{C_1^{2*}C_1^2}{i\,\Delta\omega + \mathcal{D}_{22}}\right] + \mathcal{G}_2\left[\dfrac{C_1^{1*}C_1^1}{i\,\Delta\omega + \mathcal{D}_{11}} + \dfrac{C_1^{2*}C_2^2}{i\,\Delta\omega + \mathcal{D}_{22}}\right] \\ \mathcal{G}_1\left[\dfrac{C_2^{1*}C_1^1}{i\,\Delta\omega + \mathcal{D}_{11}} + \dfrac{C_2^{2*}C_1^2}{i\,\Delta\omega + \mathcal{D}_{22}}\right] + \mathcal{G}_2\left[\dfrac{C_2^{1*}C_2^1}{i\,\Delta\omega + \mathcal{D}_{11}} + \dfrac{C_2^{2*}C_2^2}{i\,\Delta\omega + \mathcal{D}_{22}}\right] \end{bmatrix}. \tag{3-67}$$

The values for the $\mathcal{G}$'s, C's, and $\mathcal{D}$'s are calculated just once. The solution for $(\rho_{\mathrm{I}'})_{1,2}$, for instance, is obtained by evaluating Eq. (3-67) for each frequency point $\Delta\omega$ needed to describe the NMR lineshape. This procedure results in a considerable saving of computer time over the use of (3-45). First, we have reduced the size of the matrix and then diagonalization and inversion is carried out only once for each spectrum. This is in contrast with (3-45) where the A coefficient matrix has to be inverted for each frequency point.

Of course, we would not go to all these manipulations just for a single spin. The procedure only becomes worthwhile for a system of interacting spins.

One point has been consciously glossed over because it would only complicate the above discussion. In a real molecule there are usually several nuclei, with different chemical shifts and thus the $\Delta\omega$s will not all be the same. Thus, in Eq. (3-46) we could not write

$$i\,\Delta\omega\,\mathcal{G} + A_1 \tag{3-68}$$

if we were describing a system with nonequivalent spins. What we do is choose a frequency $\bar{\omega}_0$ near the center of the spectrum and define all shifts relative to it, for instance,

$$\omega - \omega_{0s} = \omega - \bar{\omega}_0 + \bar{\omega}_0 - \omega_{0s} = \Delta\bar{\omega} - \delta\bar{\omega}_s, \tag{3-69}$$

where

$$\Delta\bar{\omega} = \omega - \bar{\omega}_0 \tag{3-70}$$

and

$$\delta\bar{\omega}_s = \omega_{0s} - \bar{\omega}_0. \tag{3-71}$$

Then, the diagonal matrix $i\,\Delta\omega\,\mathcal{G}$ is replaced by $i\,\Delta\bar{\omega}\,\mathcal{G}$ and the $\delta\omega_s$ terms are in the A_1 coefficient matrix.

Summary

Using the spin $\frac{1}{2}$ problem as our model we have developed a general procedure for calculating the matrix elements of the density matrix which are needed to obtain the absorption.

Appendix

For arbitrary I

$$I^z|m\rangle = m|m\rangle \tag{3-A1}$$

$$I^+|m\rangle = \sqrt{I(I+1) - m(m+1)}\;|(m+1)\rangle \tag{3-A2}$$

$$I^-|m\rangle = \sqrt{I(I+1) - m(m-1)}\;|(m-1)\rangle \tag{3-A3}$$

Problems

1. Verify matrix (3-33).

2. Calculate $\hbar\omega_0/kT_b$ at room temperature for the proton using a 10,000-G field.

3. Using commutation rules evaluate

$$[I^+, I^z], \quad [I^-, I^z], \quad [(I^+)^2, I^-], \quad [(I^+)^2, I^z].$$

4. Evaluate

$$\mathrm{Tr}\, \rho_0 I^+ I^-, \ \mathrm{Tr}\, I^+ \rho_0 I^-$$

for

$$\rho_0 = e^{-(\hbar\omega_0 I^z/kT_b)} / \mathrm{Tr}\, e^{-(\hbar\omega_0 I^z/kT_b)},$$

where

$$\hbar\omega_0/kT \ll 1.$$

REFERENCES

1. I. Rabi, N. F. Ramsey, and J. Schwinger, *Rev. Mod. Phys.* **26**, 167 (1954).
2. R. C. Gordon and R. P. McGinnis, *J. Chem. Phys.* **49**, 2455 (1968).

Chapter IV
RELAXATION

Most NMR lineshape treatments of exchanging systems simulate relaxation effects by use of phenomenological linewidth parameters. These are included in the diagonal elements of the coefficient matrix of the density matrix equations. This is clearly unsatisfactory in all cases in which relaxation has the effect of mixing transitions, for instance, quadrupole relaxation. Then contributions from different relaxation mechanisms have to be specifically included in the density matrix equations by use of the appropriate operators; their derivation is the purpose of this chapter.

1. The Relaxation Operator

In Chapter II we derived the density matrix equation

$$\dot{\rho}_{\mathrm{T}} = -i\hbar^{-1}[\mathcal{H}_{\mathrm{T}}, \rho_{\mathrm{T}}] \tag{4-1}$$

for the relevant isolated Hamiltonian (referred to in the Introduction as the "total Hamiltonian"). In a liquid sample, assuming wall interactions are not critical, this would be the Hamiltonian of the liquid. What we will work with explicitly in this book are the Hamiltonians of *just* the spin system, $\mathcal{H}^{\mathrm{s}}$. The density matrix equation of this spin system will be of the form

$$\dot{\rho}_{\mathrm{s}} = -i\hbar^{-1}[\mathcal{H}^{\mathrm{s}}, \rho_{\mathrm{s}}] + R\rho_{\mathrm{s}}, \tag{4-2}$$

where R is the *relaxation operator* which connects the spin system to the other degrees of freedom of the system. In Chapter III R was assumed to have the form $-(\rho_s - \rho_0)/T$. Now, however, we shall derive R from the spin bath coupling interaction.

ρ_s is often called the reduced density matrix and will be differentiated from ρ_T, the "total density matrix." Note that the contributions due to exchange effects are left out in (4-2). These will be discussed in Chapter V.

In this chapter, then, we will derive the reduced density matrix for the spins in a molecule interacting with its molecular motion. The nuclear spin Hamiltonian of a molecule is

$$\mathcal{H} = \mathcal{H}_s + \mathcal{H}_{rf},$$

$$\mathcal{H}_s = \sum_s \omega_{0s} I_s^z + \sum_{s>t} J_{s,t} I_s \cdot I_t, \tag{4-3}$$

$$\mathcal{H}_{rf} = \not{\omega}_1 \sum_s \left[I_s^x \cos \omega t + I_s^y \sin \omega t \right], \tag{4-4}$$

where s, t label spins and the reduced density matrix equation will be shown to be given (for the case of random lattice magnetic fields) as

$$\frac{d\rho}{dt} = -i\hbar^{-1}\left[\mathcal{H}_s + \mathcal{H}_{rf}, \rho \right] + R\rho, \tag{4-5}$$

where

$$-R\rho = \sum_s \left\{ \frac{1}{T_{1s}} \left[I_s^+, \left[I_s^-, \tilde{\rho} - \rho_0 \right] \right] + \frac{1}{T_{1s}} \left[I_s^-, \left[I_s^+, \tilde{\rho} - \rho_0 \right] \right] + \frac{1}{T_{ts}} \left[I_s^z, \left[I_s^z, \tilde{\rho} - \rho_0 \right] \right] \right\}. \tag{4-6}$$

At the end of this chapter the definitions of T_{1s} and T_{ts} will be exhibited.

The derivation of (4-6) is long and detailed and is not essential to most further calculations in this book, but its particular simple form is the result of certain modeling approximations which should be understood by a serious student of NMR. With this proviso let us go back and derive (4-2). The derivation will be given assuming

(1) $\mathcal{H}_s$ [see (4-3)] is invariant to a rotation about the z axis, i.e.,

$$\mathcal{H}_s = e^{i\sum_s I_s^z \alpha} \mathcal{H}_s e^{-i\sum_s I_s^z \alpha} = \mathcal{H}_s. \tag{4-7}$$

(2) The applied rf field is circularly polarized.
(3) The following inequalities are satisfied

$$\omega_0 \gg J_{s,t}, \not{\omega}_1, \qquad \not{\omega}_1 \tau, J\tau \ll 1, \tag{4-8}$$

where τ is a correlation time appropriate for molecular motion.

(4) The system is in a steady state.

We start with the total density matrix equation[1]

$$\dot{\rho}_T = -i\hbar^{-1}[\mathcal{H}_T, \rho_T], \tag{4-9}$$

where

$$\mathcal{H}_T = \mathcal{H}_s + \mathcal{H}_{rf} + \mathcal{H}_b + \mathcal{H}_{sb}, \tag{4-10}$$

$\mathcal{H}_s$ and $\mathcal{H}_{rf}$ are given, respectively, in (4-3) and (4-4), $\mathcal{H}_b$ (bath) is the Hamiltonian for all other variables, and $\mathcal{H}_{sb}$ (spin bath) is the interaction connecting the bath with the spin system. Specifically, $\mathcal{H}_{sb}$ might be the dipole-dipole interaction

$$\mathcal{H}_{sb} = \mu^2\left[\frac{\mathbf{I}_1\cdot\mathbf{I}_2}{r_{1,2}^3} - \frac{3(\mathbf{I}_1\cdot\mathbf{r}_{12})(\mathbf{I}_2\cdot\mathbf{r}_{12})}{r_{1,2}^5}\right], \tag{4-11}$$

where 1, 2 are spins and $\mathbf{r}_{12}$ is the distance between them. $\mathcal{H}_{sb}$ could also be the quadrupole interaction or an anisotropic chemical shift. Further discussions of these interactions together with expressions for $\mathcal{H}_{sb}$ are given below.

We next go into the rotating coordinate system

$$\rho_T = e^{-i\omega I^z t}\tilde{\rho}_T e^{i\omega I^z t}, \tag{4-12a}$$

where

$$I^z = \sum_s I_s^z. \tag{4-12b}$$

Substituting (4-12) into (4-9) one obtains (all energies will hereafter be written in frequency units as $\mathcal{H}/\hbar$)

$$\dot{\tilde{\rho}}_T = -i[\overline{\mathcal{H}}_s + \overline{\mathcal{H}}_{rf} + \mathcal{H}_b + \overline{\mathcal{H}}_{sb}, \tilde{\rho}_T], \tag{4-13}$$

where

$$\overline{\mathcal{H}}_s = \mathcal{H}_s - \omega \sum_t I_t^z$$

and

$$\overline{\mathcal{H}}_{rf} = \omega_1 \sum_s I_s^x, \tag{4-14}$$

$$\overline{\mathcal{H}}_{sb} = e^{i\omega I^z t}\mathcal{H}_{sb} e^{-i\omega I^z t}. \tag{4-15}$$

What we want to calculate are properties of the spin systems. Thus, given a spin operator M_s its expectation value is

$$\langle M_s \rangle = \mathop{\mathrm{Tr}}_{bs} M_s \rho_T = \mathop{\mathrm{Tr}}_{bs} \overline{M}_s \tilde{\rho}_T. \tag{4-16}$$

The eigenstates of the combined system, spin s, and bath b, are written as

$$|s\rangle|b\rangle. \tag{4-17}$$

We note that M_s is only an operator in spin space. Then, $\langle M_s \rangle$ is

$$\langle M_s \rangle = \sum_{s,b} \langle sb | \overline{M}_s \tilde{\rho}_T | sb \rangle, \tag{4-18}$$

$$\langle M_s \rangle = \sum_{s,s'} \sum_{b,b'} \langle sb | \overline{M}_s | s'b' \rangle \langle s'b' | \tilde{\rho}_T | sb \rangle$$

$$= \sum_{s,s'} \sum_{b,b'} \langle b | b' \rangle \langle s | \overline{M}_s | s' \rangle \langle s'b' | \tilde{\rho}_T | sb \rangle,$$

and since

$$\langle b | b' \rangle = \delta_{b,b'} \tag{4-18a}$$

(4-18) can be rewritten

$$\langle M_s \rangle = \sum_{s,s'} \langle s | \overline{M}_s | s' \rangle \langle s' | \sum_b \langle b | \tilde{\rho}_T | b \rangle | s \rangle$$

$$= \sum_{s,s'} \langle s | \overline{M}_s | s' \rangle \langle s' | \tilde{\rho}_s | s \rangle = \mathrm{Tr}_s\, \overline{M}_s \tilde{\rho}_s = \mathrm{Tr}_s\, M_s \rho_s, \tag{4-19}$$

where

$$\tilde{\rho}_s = \sum_b \langle b | \tilde{\rho}_T | b \rangle = \mathrm{Tr}_b\, \tilde{\rho}_T\,. \tag{4-20}$$

$\tilde{\rho}_s$ is spoken of as the *reduced* density matrix and is all we need to know in order to calculate spin properties.

We extract $\tilde{\rho}_s$ from ρ_T by writing (with proper foresight)

$$\tilde{\rho}_T = \rho_0(\mathcal{H}_b)\tilde{\rho}_s + \eta(t), \tag{4-21}$$

where

$$\rho_0(\mathcal{H}_b) = \frac{e^{-\mathcal{H}_b/kT_b}}{\mathrm{Tr}\, e^{-\mathcal{H}_b/kT_b}}\,. \tag{4-22}$$

Equation (4-21) is equivalent to assuming the lattice remains in equilibrium. Recall that

$$\mathrm{Tr}\, \rho_0(\mathcal{H}_b) = 1. \tag{4-23}$$

From (4-21) and the definition of $\tilde{\rho}_s$,

$$\tilde{\rho}_s = \mathrm{Tr}_b\, \tilde{\rho}_T, \tag{4-24}$$

one obtains that

$$\eta = [1 - \rho_0(\mathcal{H}_b)\, \mathrm{Tr}_b] \rho_T = \mathcal{P} \tilde{\rho}_T. \tag{4-25}$$

We now list and prove some properties of η and $\mathcal{P}$:

(1)

$$\mathrm{Tr}_b\, \eta = 0 \tag{4-26}$$

Proof: From (4-19)

$$\begin{aligned}\tilde{\rho}_s = \mathrm{Tr}_b\, \tilde{\rho}_T &= \mathrm{Tr}_b[\rho_0(\mathcal{H}_b)\tilde{\rho}_s + \eta] \\ &= \tilde{\rho}_s\, \mathrm{Tr}_b\, \rho_0(\mathcal{H}_b) + \mathrm{Tr}_b\, \eta \\ &= \tilde{\rho}_s + \mathrm{Tr}_b\, \eta, \end{aligned} \tag{4-27}$$

$$\mathrm{Tr}_b\, \eta = 0. \tag{4-28}$$

(2)

$$\mathcal{P}\eta = \eta. \tag{4-29}$$

Proof:

$$\mathcal{P}\eta = (1 - \rho_0(\mathcal{H}_b)\,\mathrm{Tr}_b)\eta = \eta + 0 = \eta. \tag{4-30}$$

In carrying out (4-30) we have used (4-28).

(3)

$$\mathcal{P}\mathcal{H}_s - \mathcal{H}_s\mathcal{P} = [\mathcal{P}, \mathcal{H}_s] = 0. \tag{4-31}$$

This is true because $\mathcal{P}$ is solely an operator in *b* space and $\mathcal{H}_s$ solely an operator in *s* space.

(4)

$$\mathcal{P}\mathcal{P} = \mathcal{P}$$

Proof: We start with the arbitrary operator *G*.

$$\begin{aligned}\mathcal{P}\mathcal{P}G &= [1 - \rho_0(\mathcal{H}_b)\,\mathrm{Tr}_b][1 - \rho_0(\mathcal{H}_b)\,\mathrm{Tr}_b]G \\ &= G - 2\rho_0(\mathcal{H}_b)\,\mathrm{Tr}_b\, G + [\rho_0(\mathcal{H}_b)\,\mathrm{Tr}_b]G^2 \\ &= G - \rho_0(\mathcal{H}_b)\,\mathrm{Tr}_b\, G, \end{aligned} \tag{4-32}$$

where we have used the result that

$$\rho_0(\mathcal{H}_b)\,\mathrm{Tr}_b[\rho_0(\mathcal{H}_b)\,\mathrm{Tr}_b]G = \rho_0(\mathcal{H}_b)\,\mathrm{Tr}_b\, G. \tag{4-33}$$

On the other hand,

$$\mathcal{P}G = (1 - \rho_0(\mathcal{H}_b)\,\mathrm{Tr}_b)G = G - \rho_0(\mathcal{H}_b)\,\mathrm{Tr}_b\, G, \tag{4-34}$$

which is the same result as (4-32), so that

$$\mathcal{P}\mathcal{P}G = \mathcal{P}G, \tag{4-35}$$

and since *G* was chosen arbitrarily this proves that

$$\mathcal{P}\mathcal{P} = \mathcal{P}. \tag{4-36}$$

With these preliminaries aside we discuss our strategy. What we will do is obtain two coupled equations in $\tilde{\rho}_s$ and η. One of these will be solved for η in terms of $\tilde{\rho}_s$ and then substituted into the other equation to give an expression entirely in terms of $\tilde{\rho}_s$, the desired result!

To accomplish this program we make an important notational simplification, i.e., we rewrite (4-13) as

$$\dot{\tilde{\rho}}_T = -i\left(\bar{\mathcal{L}}_s + \bar{\mathcal{L}}_{rf} + \mathcal{L}_b + \bar{\mathcal{L}}_{sb}\right)\tilde{\rho}_T, \tag{4-37}$$

where

$$\mathcal{L}_i\tilde{\rho}_T = \left[\mathcal{H}_i, \tilde{\rho}_T\right]. \tag{4-38}$$

The $\mathcal{L}_i$ are called Liouville operators or superoperators. The advantage of the Liouville operator is that (4-37) now appears as a "classical" differential equation. Next, we substitute

$$\tilde{\rho}_T = \rho_0(\mathcal{H}_b)\tilde{\rho}_s + \eta \tag{4-39}$$

into (4-37) and obtain

$$\rho_0(\mathcal{H}_b)\dot{\tilde{\rho}}_s(t) + \dot{\eta}(t) = -i\left[\bar{\mathcal{L}}_s + \bar{\mathcal{L}}_{rf} + \mathcal{L}_b + \bar{\mathcal{L}}_{sb}(t)\right]\left[\rho_0(\mathcal{H}_b)\tilde{\rho}_s(t) + \eta(t)\right], \tag{4-40}$$

where here operators with an explicit time dependence are so indicated. We obtain the previously mentioned two coupled equations by operating on (4-40), first with Tr_b, obtaining

$$\dot{\tilde{\rho}}_s(t) = -i\left\{\left(\bar{\mathcal{L}}_s + \bar{\mathcal{L}}_{rf}\right)\tilde{\rho}_s(t) + \mathrm{Tr}_b\, \bar{\mathcal{L}}_{sb}(t)\rho_0(\mathcal{H}_b)\tilde{\rho}_s(t) + \mathrm{Tr}_b\, \bar{\mathcal{L}}_{sb}(t)\eta(t)\right\}, \tag{4-41}$$

and then operating on (4-40) with $\mathcal{P}$ obtaining

$$\dot{\eta}(t) = -i\left\{\left(\bar{\mathcal{L}}_s + \bar{\mathcal{L}}_{rf} + \mathcal{L}_b\right)\eta(t) + \mathcal{P}\,\bar{\mathcal{L}}_{sb}(t)\eta(t) + \mathcal{P}\,\bar{\mathcal{L}}_{sb}(t)\rho_0(\mathcal{H}_b)\tilde{\rho}_s(t)\right\}. \tag{4-42}$$

To obtain (4-41) and (4-42) we have used the results that (see problems)

$$\mathrm{Tr}_b\left(\bar{\mathcal{L}}_s + \mathcal{L}_{rf}\right)\eta = 0, \qquad \mathrm{Tr}_b\left(\mathcal{L}_b\rho_0(\mathcal{H}_b)\right)\tilde{\rho}_s = 0, \qquad \mathrm{Tr}_b\, \mathcal{L}_b\eta = 0, \tag{4-43}$$

$$\mathcal{P}\left(\bar{\mathcal{L}}_s + \mathcal{L}_b + \bar{\mathcal{L}}_{rf}\right)\rho_0(\mathcal{H}_b)\tilde{\rho}_s = 0,$$

$$\mathcal{P}\left(\bar{\mathcal{L}}_s + \mathcal{L}_b + \bar{\mathcal{L}}_{rf}\right)\eta = \left(\bar{\mathcal{L}}_s + \mathcal{L}_b + \bar{\mathcal{L}}_{rf}\right)\eta.$$

These relations are proved using primarily the result proved in Appendix A that

$$\mathrm{Tr}_b\left[A(b), B(b, s)\right] = 0, \tag{4-44}$$

where $A(b)$ is solely an operator in b space and $B(b, s)$ an operator in b and s space.

As suggested a few paragraphs back, we next solve (4-42) for $\eta(t)$ as a function of $\tilde{\rho}_s(t)$. Before doing this we can simplify the calculation by noting that as we will only want to calculate the relaxation operator R to second order in $\mathcal{H}_{sb}$ (we assume a weak coupling to the lattice), we can drop $\mathcal{P}\,\mathcal{L}_{sb}(t)\eta(t)$ on the right-hand side of Eq. (4-42). This will be justified a few steps further on. Thus, we are left with

$$\eta(t) = -i\Big\{\big(\bar{\mathcal{L}}_s + \bar{\mathcal{L}}_{rf} + \mathcal{L}_b\big)\eta(t) + \mathcal{P}\,\bar{\mathcal{L}}_{sb}\rho_0(\mathcal{H}_b)\tilde{\rho}_s(t)\Big\}, \tag{4-45}$$

which for the initial condition at $t = 0$ that [see (4-25)]

$$\eta(0) = \mathcal{P}\rho_0\big(\bar{\mathcal{H}}_s + \mathcal{H}_b\big) \tag{4-46}$$

has the solution

$$\eta(t) = e^{-i(\bar{\mathcal{L}}_s + \bar{\mathcal{L}}_{rf} + \mathcal{L}_b)t}\mathcal{P}\rho_0(\mathcal{H}_s + \mathcal{H}_b) + \int_0^t e^{-i(\bar{\mathcal{L}}_s + \bar{\mathcal{L}}_{rf} + \mathcal{L}_b)(t-t')} \times\Big[-i\mathcal{P}\,\bar{\mathcal{L}}_{sb}(t')\rho_0(\mathcal{H}_b)\tilde{\rho}_s(t')\Big]\,dt'. \tag{4-47}$$

We note that

$$\begin{aligned}\mathcal{P}\rho_0\big(\bar{\mathcal{H}}_s + \mathcal{H}_b\big) &= \mathcal{P}\rho_0\big(\bar{\mathcal{H}}_s\big)\rho_0(\mathcal{H}_b) \\ &= \rho_0\big(\bar{\mathcal{H}}_s\big)(1 - \rho_0(\mathcal{H}_b)\,\mathrm{Tr}_b)\rho_0(\mathcal{H}_b) \\ &= \rho_0\big(\bar{\mathcal{H}}_s\big)(\rho_0(\mathcal{H}_b) - \rho_0(\mathcal{H}_b)) = 0,\end{aligned} \tag{4-48}$$

which eliminates the first term on the right-hand side of (4-47). Substituting (4-47) (with the first term omitted) into (4-41) yields

$$\dot{\tilde{\rho}}_s = -i\big(\bar{\mathcal{L}}_s + \bar{\mathcal{L}}_{rf}\big)\tilde{\rho}_s - i\,\mathrm{Tr}_b\,\bar{\mathcal{L}}_{sb}\rho_0(\mathcal{H}_b)\tilde{\rho}_s(t) - \mathrm{Tr}_b\,\bar{\mathcal{L}}_s(t)\int_0^t e^{-i(\bar{\mathcal{L}}_s + \bar{\mathcal{L}}_{rf} + \mathcal{L}_b)(t-t')}\mathcal{P}\,\bar{\mathcal{L}}_{sb}(t')\rho_0(\mathcal{H}_b)\tilde{\rho}_s(t')\,dt'. \tag{4-49}$$

The first term in (4-49) represents the interaction of the isolated spin system. The second term is a first-order interaction with $\mathcal{H}_{sb}$. This will be zero when the interaction, as we will assume, is off diagonal in the bath coordinates. The last term in (4-49) gives rise to the coupling of the spin system to the lattice. Note that it is second order in $\mathcal{H}_{sb}$ which justifies our setting

$$\mathcal{P}\,\bar{\mathcal{L}}_{sb}(t)\eta(t) = 0 \tag{4-50}$$

in (4-42). Also notice that aside from keeping $\mathcal{H}_{sb}$ only to second order the theory is exact.

We now define

$$\tau = t - t', \tag{4-51}$$

which changes (4-49) to

$$\begin{aligned}\dot{\tilde{\rho}}_s(t) = &-i(\bar{\mathcal{L}}_s + \bar{\mathcal{L}}_{rf})\tilde{\rho}_s(t)\\ &-\mathrm{Tr}_b\, \bar{\mathcal{L}}_{sb}(t)\int_0^t e^{-i(\bar{\mathcal{L}}_s + \bar{\mathcal{L}}_{rf} + \mathcal{L}_b)\tau}\\ &\times \mathcal{P}\, \bar{\mathcal{L}}_{sb}(t-\tau)\rho_0(\mathcal{H}_b)\tilde{\rho}_s(t-\tau)\, d\tau.\end{aligned} \tag{4-52}$$

Now, one can see the advantage of being in the rotating coordinate system (i.e., $\overline{\mathcal{H}}_s$ and $\overline{\mathcal{H}}_{rf}$ are independent of time), for after turning on the rf field at $t = 0$, $\tilde{\rho}_s(t)$ will have an initial transient time dependence, and then after some time ($t \to \infty$) will settle down to a constant value. Thus, for very long times (4-52) becomes

$$\begin{aligned}\dot{\tilde{\rho}}_s(t) = 0 = &-i(\bar{\mathcal{L}}_s + \bar{\mathcal{L}}_{rf})\tilde{\rho}_s - \mathrm{Tr}_b\, \bar{\mathcal{L}}_{sb}(t)\int_0^\infty e^{-i(\bar{\mathcal{L}}_s + \bar{\mathcal{L}}_{rf} + \mathcal{L}_b)\tau}\\ &\times \bar{\mathcal{L}}_{sb}(t-\tau)\rho_0(\mathcal{H}_b)\tilde{\rho}_s\, d\tau.\end{aligned} \tag{4-53}$$

Note that in going from (4-52) to (4-53) we have used the result that

$$\mathcal{P}\, \bar{\mathcal{L}}_{sb}(t-\tau)\rho_0(\mathcal{H}_b)\tilde{\rho}_s = \bar{\mathcal{L}}_{sb}(t-\tau)\rho_0(\mathcal{H}_b)\tilde{\rho}_s\, . \tag{4-54}$$

Proof:

$$\begin{aligned}&\mathcal{P}\, \bar{\mathcal{L}}_{sb}(t-\tau)\rho_0(\mathcal{H}_b)\rho_s\\ &\quad= (1 - \rho_0(\mathcal{H}_b)\,\mathrm{Tr}_b)\left\{\overline{\mathcal{H}}_{sb}(t-\tau)\rho_0(\mathcal{H}_b)\tilde{\rho}_s - \rho_0(\mathcal{H}_b)\tilde{\rho}_s\overline{\mathcal{H}}_{sb}(t-\tau)\right\}\\ &\quad= \bar{\mathcal{L}}_{sb}(t-\tau)\rho_0(\mathcal{H}_b)\tilde{\rho}_s\\ &\quad\quad -\rho_0(\mathcal{H}_b)\sum_b \left\{\langle b|\overline{\mathcal{H}}_{sb}(t-\tau)|b\rangle\rho_0(E_b)\tilde{\rho}_s\right.\\ &\quad\quad \left. -\rho_0(E_b)\tilde{\rho}_s\langle b|\overline{\mathcal{H}}_{sb}(t-\tau)|b\rangle\right\},\end{aligned} \tag{4-55}$$

where b is the representation in which $\mathcal{H}_b$ is diagonal. Now,

$$\begin{aligned}\langle b|\overline{\mathcal{H}}_{sb}(t-\tau)|b\rangle &= \langle b|e^{i\omega I^z(t-\tau)}\mathcal{H}_{sb}e^{-i\omega I^z(t-\tau)}|b\rangle\\ &= e^{i\omega I^z(t-\tau)}\langle b|\mathcal{H}_{sb}|b\rangle e^{-i\omega I^z(t-\tau)}\\ &= 0,\end{aligned} \tag{4-56}$$

as we have previously assumed $\langle b|\mathcal{H}_{sb}|b\rangle = 0$ in a representation in which $\mathcal{H}_b$ is diagonal. Thus (4-54) is proved.

Equation (4-53) is our final formal result. The next step is to express quite generally that[2-4]

$$\mathcal{H}_{sb} = \sum_{i,\alpha} \mathcal{G}_i^\alpha \nu_i^\alpha, \tag{4-57}$$

where $\mathcal{G}$'s are solely spin operators and ν's solely lattice operators; spins are labeled i. The $\mathcal{G}$'s are defined so that

$$e^{i\omega I'_z t} \mathcal{G}_i^\alpha e^{-i\omega I'_z t} = e^{i\omega_\alpha t} \mathcal{G}_i^\alpha. \tag{4-58}$$

Every nuclear spin relaxation mechanism has its own $\mathcal{H}_{sb}$ operator. Thus, for intramolecular dipole relaxation (spins within the same molecule), $\mathcal{H}_{sb}$ is

$$\mathcal{H}_{sb} = \sum_{i,j} \mu^2 \left[\frac{\mathbf{I}_i \cdot \mathbf{I}_j}{r_{i,j}^3} - \frac{3(\mathbf{I}_i \cdot \mathbf{r}_{i,j}{}^2)(\mathbf{I}_i \cdot \mathbf{r}_{i,j}^{\ 2})}{r_{i,j}^5} \right], \tag{4-59}$$

where interactions between all pairs of spins are summed and the $\mathbf{r}_{ij}$ vectors connect spins i and j. We now list the relevant operators for intramolecular dipole-dipole relaxation, considering two spins labeled 1 and 2.[§]

$$\alpha = 0, \pm 1, \pm 2, \tag{4-60}$$

$$\mathcal{G}_{1,2}^0 = [I_1^z I_2^z - \tfrac{1}{4}(I_1^+ I_2^- + I_1^- I_2^+)], \qquad \nu_{1,2}^0 = k_{1,2}(1 - 3\cos^2\theta),$$

$$\mathcal{G}_{1,2}^1 = (I_1^+ I_2^- + I_1^- I_2^+), \qquad \nu_{1,2}^1 = -\tfrac{3}{2} k_{1,2} \sin\theta \cos\theta\, e^{i\varphi},$$

$$\mathcal{G}_{1,2}^{-1} = (\mathcal{G}_{1,2}^1)^\dagger, \qquad \nu_{1,2}^{-1} = (\nu_{1,2}^1)^*,$$

$$\mathcal{G}_{1,2}^2 = I_1^+ I_2^+, \qquad \nu_{1,2}^2 = -\tfrac{3}{4} k_{1,2} \sin^2\theta\, e^{i2\varphi},$$

$$\mathcal{G}_{1,2}^{-2} = (\mathcal{G}_{1,2}^2)^\dagger, \qquad \nu_{1,2}^{-2} = \nu_{1,2}^2{}^*,$$

$$k_{i,j} = \hbar\gamma_1\gamma_2/r_{1,2}^3.$$

For a spin within a molecule interacting via the dipolar interaction with spins external to the molecule, the terms in $\mathcal{H}_{sb}$ can be sometimes represented as

$$\mathcal{H}_{sb} = \sum_{i,\alpha} \mathcal{G}_i^\alpha \nu_i^\alpha, \qquad \alpha = 0, \pm 1,$$

$$\mathcal{G}_i^0 = I_i^z, \qquad \mathcal{G}_i^{+1} = I_i^+, \qquad \mathcal{G}_i^{-1} = I_i^-, \tag{4-61}$$

and the νs can be thought of as random magnetic fields. This will be true

---

[§]Where θ, ϕ are the spherical coordinate angles of the vector connecting spin 1 to spin 2 relative to the laboratory x, y, z axes.

only in the limit of low power [see (4-190)]. The form given in (4-61) can also be shown to result from anisotropy of the chemical shift.

In the rotating coordinate system

$$\overline{\mathcal{H}}_{\text{sb}}(t) = e^{i\omega I^z t} \sum_i \sum_\alpha \mathcal{G}_i^\alpha v_i^\alpha e^{-i\omega I^z t}$$

$$= \sum_i \sum_\alpha e^{i\omega\alpha t} \mathcal{G}_i^\alpha v_i^\alpha$$

$$= \sum_{i,\alpha} \mathcal{H}_i^\alpha e^{i\omega\alpha t} \tag{4-62a}$$

and

$$\overline{\mathcal{L}}_{\text{sb}}(t)A = \left[\overline{\mathcal{H}}_{\text{sb}}(t), \mathrm{A}\right] = \sum_{i,\alpha} e^{i\omega\alpha t} \mathcal{L}_{i,\alpha} \mathrm{A}. \tag{4-62b}$$

Substituting (4-62) into (4-53) and recalling the definition of $\mathcal{L}$ [(4-38)] one has

$$0 = -i\left(\overline{\mathcal{L}}_{\text{s}} + \overline{\mathcal{L}}_{\text{rf}}\right)\tilde{\rho}_{\text{s}} + R\tilde{\rho}_{\text{s}}, \tag{4-63}$$

where

$$-R\tilde{\rho}_{\text{s}} = \mathrm{Tr}_{\text{b}} \sum_{i,j} \sum_{i',\alpha'} e^{i\omega\alpha t} e^{i\omega\alpha' t} \mathcal{L}_{i,\alpha}$$

$$\times \int_0^\infty e^{-i(\overline{\mathcal{L}}_{\text{s}} + \mathcal{L}_{\text{b}} + \overline{\mathcal{L}}_{\text{rf}})\tau} e^{-i\omega\alpha'\tau}$$

$$\times \mathcal{L}_{i',\alpha'} \rho_0(\mathcal{H}_{\text{b}})\tilde{\rho}_{\text{s}}\, d\tau. \tag{4-64}$$

Now, as we previously argued that $\tilde{\rho}_{\text{s}}$ is independent of the time (we set $\dot{\tilde{\rho}} = 0$), it follows that $R\tilde{\rho}_{\text{s}}$ in (4-64) must also be time independent, which in turn implies that

$$e^{i\omega\alpha t} e^{i\omega\alpha' t} = 1 \tag{4-65}$$

or

$$\alpha = -\alpha'. \tag{4-66}$$

Using (4-66), Eq. (4-64) simplifies to

$$-R\tilde{\rho}_{\text{s}} = \mathrm{Tr}_{\text{b}} \sum_{i,i'} \sum_\alpha \mathcal{L}_{i,\alpha} \int_0^\infty e^{-i(\overline{\mathcal{L}}_{\text{s}} + \mathcal{L}_{\text{b}} + \overline{\mathcal{L}}_{\text{rf}})\tau}\, e^{i\omega\alpha\tau} \mathcal{L}_{i,-\alpha} \rho_0(\mathcal{H}_{\text{b}})\tilde{\rho}_{\text{s}}\, d\tau. \tag{4-67}$$

Also, using the definition

$$\mathcal{L}_i \mathrm{A} = [\mathcal{H}_i, \mathrm{A}] \tag{4-68}$$

and the relationship that

$$e^{i\mathcal{L}_i\tau}\mathrm{A} = \mathrm{A} + i\tau\mathcal{L}A - \frac{1}{2!}\tau^2\mathcal{L}_i\mathcal{L}_i\mathrm{A}$$
$$= \mathrm{A} + i\tau[\mathcal{H}_i\,,\mathrm{A}] - \frac{1}{2!}[\mathcal{H}_i\,,[\mathcal{H}_i\,,\mathrm{A}]]$$
$$= e^{i\mathcal{H}_i\tau}\mathrm{A}e^{-i\mathcal{H}_i\tau}, \tag{4-69}$$

we can express (4-67) as

$$-R\tilde{\rho}_{\mathrm{s}} = \mathrm{Tr}_{\mathrm{b}}\sum_{i,\,i'}\sum_{\alpha}\Bigg[\mathcal{H}_{i,\,\alpha}\,,\int_0^{\infty} e^{i\omega\alpha\tau}e^{-i(\overline{\mathcal{H}}_{\mathrm{s}}+\mathcal{H}_{\mathrm{b}}+\overline{\mathcal{H}}_{\mathrm{rf}})\tau}$$
$$\times[\mathcal{H}_{i',\,-\alpha}\,,\rho_0(\mathcal{H}_{\mathrm{b}})\tilde{\rho}_{\mathrm{s}}]e^{i(\overline{\mathcal{H}}_{\mathrm{s}}+\mathcal{H}_{\mathrm{b}}+\overline{\mathcal{H}}_{\mathrm{rf}})\tau}\Bigg]\,d\tau. \tag{4-70}$$

Expanding the commutator and defining

$$\nu_i^{\alpha}(\tau) = e^{i\mathcal{H}_{\mathrm{b}}\tau}\nu_i^{\alpha}e^{-i\mathcal{H}_{\mathrm{b}}\tau} \tag{4-71}$$

one obtains

$$R\rho =$$
$$-\sum_{i,\,i'}\sum_{\alpha}\mathrm{Tr}_{\mathrm{b}}\Bigg\{\int_0^{\infty}\mathcal{G}_i^{\alpha}\nu_i^{\alpha}e^{i\omega\alpha\tau}e^{-i(\overline{\mathcal{H}}_{\mathrm{s}}+\overline{\mathcal{H}}_{\mathrm{rf}})\tau}\mathcal{G}^{-\alpha}\nu_{i'}^{-\alpha}(-\tau)\rho_0(\mathcal{H}_{\mathrm{b}})\rho_{\mathrm{s}}e^{i(\overline{\mathcal{H}}_{\mathrm{s}}+\overline{\mathcal{H}}_{\mathrm{rf}})\tau}\,d\tau$$
$$-\int_0^{\infty}\mathcal{G}_i^{\alpha}\nu_i^{\alpha}e^{i\omega\alpha\tau}e^{-i(\overline{\mathcal{H}}_{\mathrm{s}}+\overline{\mathcal{H}}_{\mathrm{rf}})\tau}\rho_0(\mathcal{H}_{\mathrm{b}})\rho_{\mathrm{s}}\mathcal{G}_{i'}^{-\alpha}e^{i(\overline{\mathcal{H}}_{\mathrm{s}}+\overline{\mathcal{H}}_{\mathrm{rf}})\tau}\nu_{i'}^{-\alpha}(-\tau)\,d\tau$$
$$-\int_0^{\infty}e^{i\omega\alpha\tau}e^{-(\overline{\mathcal{H}}_{\mathrm{s}}+\overline{\mathcal{H}}_{\mathrm{rf}})\tau}\mathcal{G}_{i'}^{-\alpha}\nu_{i'}^{-\alpha}(-\tau)\rho_0(\mathcal{H}_{\mathrm{b}})\rho_{\mathrm{s}}e^{i(\overline{\mathcal{H}}_{\mathrm{s}}+\overline{\mathcal{H}}_{\mathrm{rf}})\tau}\mathcal{G}_i^{\alpha}\nu_i^{\alpha}\,d\tau$$
$$+\int_0^{\infty}e^{i\omega\alpha\tau}e^{-i(\overline{\mathcal{H}}_{\mathrm{s}}+\overline{\mathcal{H}}_{\mathrm{rf}})\tau}\rho_0(\mathcal{H}_{\mathrm{b}})\rho_{\mathrm{s}}\mathcal{G}_{i'}^{-\alpha}e^{i(\overline{\mathcal{H}}_{\mathrm{s}}+\overline{\mathcal{H}}_{\mathrm{rf}})\tau}\nu_{i'}^{-\alpha}(-\tau)\mathcal{G}_i^{\alpha}\nu_i^{\alpha}\,d\tau\Bigg\}. \tag{4-72}$$

We next define the correlation functions of the bath variables as

$$c_{\alpha(i,\,i')}(\tau) = \mathrm{Tr}_{\mathrm{b}}\,\rho_0(\mathcal{H}_{\mathrm{b}})\nu_i^{\alpha}(\tau)\nu_{i'}^{-\alpha} = \mathrm{Tr}_{\mathrm{b}}\,\rho_0(\mathcal{H}_{\mathrm{b}})\nu_i^{\alpha}(t+\tau)\nu_{i'}^{-\alpha}(t). \tag{4-73}$$

In terms of $c_{\alpha(i,\,i')}(\tau)$ (4-72) can be rewritten as

$$-R\tilde{\rho}_{\mathrm{s}} = \sum_{\alpha(i,\,i')}\sum\int_0^{\infty}d\tau\,e^{i\omega\alpha\tau}c_{\alpha(i,\,i')}(\tau)\Big[\mathcal{G}_i^{\alpha}\,,e^{-i(\overline{\mathcal{H}}_{\mathrm{s}}+\overline{\mathcal{H}}_{\mathrm{rf}})\tau}\mathcal{G}_{i'}^{-\alpha}\tilde{\rho}_{\mathrm{s}}e^{i(\overline{\mathcal{H}}_{\mathrm{s}}+\overline{\mathcal{H}}_{\mathrm{rf}})\tau}\Big]$$
$$-\sum_{\alpha(i,\,i')}\sum\int_0^{\infty}d\tau\,e^{i\omega\alpha\tau}c_{-\alpha(i',\,i)}(-\tau)\Big[\mathcal{G}_i^{\alpha}\,,e^{-i(\overline{\mathcal{H}}_{\mathrm{s}}+\overline{\mathcal{H}}_{\mathrm{rf}})\tau}\tilde{\rho}_{\mathrm{s}}\mathcal{G}_{i'}^{-\alpha}e^{i(\overline{\mathcal{H}}_{\mathrm{s}}+\overline{\mathcal{H}}_{\mathrm{rf}})\tau}\Big]. \tag{4-74}$$

To motivate what we are going to do next consider a simple two level nuclear spin system with populations n_1 and n_2 and energies E_1 and E_2. The rate equations for this system are

$$dn_1/dt = -w_{12}n_1 + w_{21}n_2, \qquad dn_2/dt = -w_{21}n_2 + w_{12}n_1. \tag{4-75}$$

At equilibrium $dn/dt = 0$, which from Eq. (4-75) implies that

$$w_{12}/w_{21} = n_2/n_1 = e^{-\beta(E_2-E_1)}. \tag{4-76}$$

The cs in (4-74) contain the temperature dependence of the relaxation rate, and, thus, since our system must be such that it attempts to return to equilibrium, we expect a similar relation to exist among the cs as that between ω_{12} and ω_{21}. It turns out that to show this relationship we must define

$$c^{\pm}_{\alpha(i,i')}(\tau) = \tfrac{1}{2}\left[c_{\alpha(i,i')}(+\tau) \pm c_{-\alpha(i',i)}(-\tau)\right]. \tag{4-77}$$

With this change in notation we can rewrite (4-74) as

$$-R\tilde{\rho}_s = \sum_\alpha \sum_{i,i'} \int_0^\infty d\tau\, e^{i\omega\alpha\tau}\Big\{c^{+}_{\alpha(i,i')}(\tau)\Big[\mathcal{G}_i^\alpha, e^{-i(\mathcal{H}_s+\mathcal{H}_{rf})\tau}\big[\mathcal{G}_{i'}^{-\alpha}, \tilde{\rho}_s\big]e^{i(\mathcal{H}_s+\mathcal{H}_{rf})\tau}\Big]$$
$$+ c^{-}_{\alpha(i,i')}(\tau)\Big[\mathcal{G}_i^\alpha, e^{-i(\mathcal{H}_s+\mathcal{H}_{rf})\tau}\big[\mathcal{G}_{i'}^{-\alpha}, \tilde{\rho}_s\big]_+ e^{i(\mathcal{H}_s+\mathcal{H}_{rf})\tau}\Big]\Big\}, \tag{4-78}$$

where

$$[A,B] = AB - BA, \qquad [A,B]_+ = AB + BA. \tag{4-79}$$

At this point we will derive the temperature connection between the $c^{\pm}$s. First, define[5, 6]

$$j^{\pm}(\omega) = \frac{1}{2}\int_{-\infty}^{\infty} d\tau\, e^{i\omega\tau} c^{\pm}(\tau). \tag{4-80}$$

$c^{\pm}(\tau)$ are to be thought of as correlation functions, with a qualitative behavior given as

$$c^{\pm}(\tau) = c^{\pm}(0)e^{-|\tau|/\tau_c}, \tag{4-81}$$

where τ_c is the correlation time. Thus, in (4-78) we take $c^{\pm}(\tau)$ to be even in τ, i.e.,

$$c^{\pm}(\tau) = c^{\pm}(-\tau), \tag{4-82}$$

and in addition we assume $c^{\pm}(\tau)$ to be real. Allowing for $c^{\pm}(\tau)$ being imaginary would give rise to small frequency shifts which we assume have been previously absorbed in $\mathcal{H}_s$. Next, invert (4-80) to obtain

$$c^{\pm}(\tau) = \frac{1}{\pi}\int_{-\infty}^{+\infty} j^{\pm}(\omega)e^{-i\omega\tau}\, d\omega. \tag{4-83}$$

Evaluating $c^{\pm}(\tau)$, see (4-73) and (4-77), in a representation in which $\mathcal{H}_b$ is diagonal one has

$$c^{\pm}_{\alpha(i,i')}(\tau) = \tfrac{1}{2} \sum_{b,b'} \langle b| v_i^{\alpha} |b'\rangle\langle b'| v_{i'}^{-\alpha} |b\rangle e^{-i(E_b - E_{b'})\tau}\left[e^{-E_b\beta} \pm e^{-E_{b'}\beta}\right], \tag{4-84}$$

where

$$\beta = 1/kT_b .$$

Then, defining

$$G_{\alpha(i,i')}(\omega) = \frac{1}{\pi} \sum_{b,b'} \langle b|v_i^{\alpha}|b'\rangle\langle b'|v_{i'}^{-\alpha}|b\rangle e^{-E_b\beta}\delta(E_{b'} - E_b - \omega), \tag{4-85}$$

one sees that (4-83) will be satisfied by setting

$$j^{\pm}_{\alpha(i,i')}(\omega) = G_{\alpha(i,i')}(\omega)\left[1 \pm e^{-\beta\omega}\right]. \tag{4-86}$$

Therefore, one has

$$\frac{j^{+}(\omega)}{j^{-}(\omega)} = \frac{1 + e^{-\beta\omega}}{1 - e^{-\beta\omega}} = \frac{1}{\tanh\frac{1}{2}\beta\omega} \tag{4-87}$$

or

$$j^{-}(\alpha\omega) = j^{+}(\alpha\omega) \tanh\tfrac{1}{2}\alpha\beta\omega, \tag{4-88}$$

where for generality we have replaced ω by $\alpha\omega$. Using (4-80) we rewrite (4-78) as

$$R\tilde{\rho}_s = -\sum_{\alpha(i,i')}\sum \Big\{\Big[\mathcal{G}_i^{\alpha}, j^{+}_{\alpha(i,i')}\big(\alpha\omega - \bar{\mathcal{L}}_s - \bar{\mathcal{L}}_{rf}\big)\big[\mathcal{G}_{i'}^{-\alpha}, \tilde{\rho}_s\big]\Big]$$
$$+ \Big[\mathcal{G}_i^{\alpha}, j^{-}_{\alpha(i,i')}\big(\alpha\omega - \bar{\mathcal{L}}_s - \bar{\mathcal{L}}_{rf}\big)\big[\mathcal{G}_{i'}^{-\alpha}, \tilde{\rho}_s\big]_{+}\Big]\Big\}, \tag{4-89}$$

where

$$j^{\pm}_{\alpha(i,i')}(\alpha\omega - \bar{\mathcal{L}})\mathrm{A} = \tfrac{1}{2}\int_{-\infty}^{+\infty} d\tau\, e^{i\alpha\omega\tau} c^{\pm}_{\alpha(i,i')}(\tau) e^{-i\mathcal{H}\tau}\mathrm{A}e^{i\mathcal{H}\tau}. \tag{4-90}$$

Equation (4-89) is our final general result.

2. Relaxation Mechanisms[7–10]

a. Random Field Interaction

To proceed further we must choose a relaxation mechanism and give $\mathcal{H}_{sb}$ the appropriate form: For simplicity we choose the simplest form of $\mathcal{H}_{sb}$, i.e., the random field interaction, which was previously given in

(4-61) as

$$\alpha = 0, \pm 1, \tag{4-91}$$

$$\mathcal{G}_i^0 = I_i^z, \qquad \mathcal{G}_i^{\pm 1} = I_i^{\pm}. \tag{4-92}$$

In addition, we have previously assumed that the correlation times of $c(\tau)$ are short enough such that

$$\not\omega_1 \tau_c, J\tau_c \ll 1. \tag{4-8}$$

Thus, one can neglect $J\mathbf{I}_i \cdot \mathbf{I}_j$ and $\overline{\mathcal{H}}_{rf}$ where they appear in (4-78) and (4-89) in the term

$$e^{i(\overline{\mathcal{H}}_s + \overline{\mathcal{H}}_{rf})\tau}, \tag{4-93}$$

which leaves just

$$e^{i\overline{\mathcal{H}}_s^0 \tau} = e^{i\Sigma_i(\omega_{0i} - \omega)I^z\tau}, \tag{4-94}$$

where $\mathcal{H}_s^0$ contains only the chemical shift terms.

Thus, (4-89) can now be written as

$$R\tilde{\rho}_s = - \sum_{\alpha(i,i')} \Big\{ \Big[\mathcal{G}_i^\alpha, j^+_{\alpha(i,i')}\big(\omega\alpha - \overline{\mathcal{L}}_s^0\big)\big[\mathcal{G}_i^{-\alpha}, \tilde{\rho}_s\big]\Big] + \Big[\mathcal{G}_i^\alpha, j^-_{\alpha(i,i')}\big(\omega\alpha - \overline{\mathcal{L}}_s^0\big)\big[\mathcal{G}_i^{-\alpha}, \tilde{\rho}_s\big]_+\Big]. \tag{4-95}$$

We now further simplify (4-95) by retaining only terms with $i = i'$. The rationale for this lies in assuming that the random fields acting on spins i and j are uncorrelated. With this assumption (4-95) now appears as

$$- R\tilde{\rho}_s = \sum_{\alpha,i} \Big\{ \Big[\mathcal{G}_i^\alpha, j^+_{\alpha,i}\big(\omega\alpha - \overline{\mathcal{L}}_s^0\big)\big[\mathcal{G}_i^{-\alpha}, \tilde{\rho}_s\big]\Big] + \Big[\mathcal{G}_i^\alpha, j^-_{\alpha,i}\big(\omega\alpha - \overline{\mathcal{L}}_s^0\big)\big[\mathcal{G}_i^{-\alpha}, \tilde{\rho}_s\big]_+\Big]. \tag{4-96}$$

We will next show that to order $(kT)^{-1}$ (4-96) is equivalent to

$$- R\tilde{\rho}_s = \sum_{\alpha,i} \Big[\mathcal{G}_i^\alpha, j^+_{\alpha,i}\big(\alpha\omega - \overline{\mathcal{L}}_s^0\big)\big[\mathcal{G}_i^{-\alpha}, \tilde{\rho}_s - \rho_0\big]\Big] \tag{4-97}$$

and under suitable conditions to

$$- R\tilde{\rho}_s = \sum_{\alpha,i} j_i^+(\alpha\omega_{0i})\big[\mathcal{G}_i^\alpha, \big[\mathcal{G}_i^{-\alpha}, \tilde{\rho}_s - \rho_0\big]\big], \tag{4-98}$$

which for the random field interaction may be rewritten as

$$- R\tilde{\rho}_s = \sum_i \Big\{ \frac{1}{T_{ti}}\big[I_i^z, \big[I_i^z, \tilde{\rho}_s - \rho_0\big]\big] + \frac{1}{T_{1i}}\big[I_i^+, \big[I_i^-, \tilde{\rho}_s - \rho_0\big]\big] + \frac{1}{T_{1i}}\big[I_i^-, \big[I_i^+, \tilde{\rho}_s - \rho_0\big]\big]\Big\}. \tag{4-99}$$

In (4-97)–(4-99) we are using the equilibrium density matrix

$$\rho_0 = e^{-\hbar\beta\omega_0 I^z}/\mathrm{Tr}\, e^{-\hbar\beta\omega_0 I^z} \tag{4-100}$$

in the form

$$\rho_0 \simeq (1/N)(1 - \hbar\beta\omega_0 I^z).$$

Now, let us go back and prove how to go from (4-96) to (4-97), which is equivalent to showing how

$$j^-\left(\alpha\omega - \bar{\mathfrak{L}}_s^0\right)\left[\mathcal{I}^{-\alpha}, \tilde{\rho}_s\right]_+ \tag{4-101}$$

equals

$$-j^+\left(\alpha\omega - \bar{\mathfrak{L}}_s^0\right)\left[\mathcal{I}^{-\alpha}, \rho_0\right]. \tag{4-102}$$

In the high temperature limit using

$$\tilde{\rho}_s \simeq 1/N + \rho_1 \tag{4-103}$$

one sees that

$$\left[\mathcal{I}^{-\alpha}, \tilde{\rho}_s\right]_+ \simeq 2\mathcal{I}^{-\alpha}/N. \tag{4-104}$$

Also as

$$j^-\left(\alpha\omega - \bar{\mathfrak{L}}_s\right) = \tanh\tfrac{1}{2}\alpha\hbar\omega\beta j^+\left(\alpha\omega - \bar{\mathfrak{L}}_s^0\right) \tag{4-105}$$

$$\simeq (\alpha\hbar\omega\beta/2) j^+\left(\alpha\omega - \bar{\mathfrak{L}}_s^0\right), \tag{4-106}$$

then (4-101) can be rewritten as

$$j^-\left(\alpha\omega - \bar{\mathfrak{L}}_s^0\right)\left[\mathcal{I}^{-\alpha}, \tilde{\rho}_s\right]_+ = (\alpha\hbar\omega\beta/N) j^+\left(\alpha\omega - \bar{\mathfrak{L}}_s^0\right)\mathcal{I}^{-\alpha}. \tag{4-107}$$

Also, in the high temperature limit

$$\rho_0 \simeq (1 - \hbar\omega_0\beta I^z)/N, \tag{4-108}$$

which converts to (4-102),

$$-j^+\left(\alpha\omega - \bar{\mathfrak{L}}_s\right)\left[\mathcal{I}^{-\alpha}, \rho_0\right] \sim (\hbar\omega_0\beta/N) j^+\left(\alpha\omega - \bar{\mathfrak{L}}_s^0\right)\left[\mathcal{I}^{-\alpha}, I^z\right]. \tag{4-109}$$

From the definition of $\mathcal{I}^\alpha$,

$$e^{i\omega I^z\tau}\mathcal{I}^\alpha e^{-i\omega I^z\tau} = e^{i\alpha\omega\tau}\mathcal{I}^\alpha, \tag{4-110}$$

it follows that

$$\left[I^z, \mathcal{I}^\alpha\right] = \alpha\mathcal{I}^\alpha, \tag{4-111}$$

and so (4-109) may be written as

$$-j^+\left(\alpha\omega - \bar{\mathfrak{L}}_s^0\right)\left[\mathcal{I}^{-\alpha}, \rho_0\right] = (\alpha\hbar\omega\beta/N) j^+\left(\alpha\omega - \bar{\mathfrak{L}}_s^0\right)\mathcal{I}^{-\alpha}, \tag{4-112}$$

and noting the near equivalence $\omega \simeq \omega_0$ over the linewidth of an absorption completes the exhibition of the equivalence of (4-101) and (4-102) and explains how one proceeds from (4-96) to (4-97).

At this point it is useful to examine the different levels of approximation at which the relaxation operator R may be written. Equation (4-97) arises principally from the high temperature approximation.

$$\hbar\omega \ll kT. \tag{4-113}$$

If one is dealing with NMR response close to resonance, $\omega_0 - \omega$ will be small (say 10^4 rad $\sec^{-1}$), so that the exponential operators in j^+ can be approximated as

$$e^{i\alpha\omega\tau} e^{\pm i(\omega_0 - \omega) I^z \tau} \simeq e^{i\alpha\omega\tau} \tag{4-114}$$

if $(\omega_0 - \omega)\tau_c \ll 1$, where τ_c is the correlation time of molecular motion. Then,

$$j^+\left(\alpha\omega - \bar{\mathfrak{L}}_s\right) \tag{4-115}$$

becomes approximately

$$j^+(\alpha\omega_0) = \tfrac{1}{2}\int_{-\infty}^{+\infty} c_\alpha^+(\tau) e^{i\alpha\omega\tau}\, d\tau \tag{4-116}$$

and the relaxation operator is given as

$$-R\tilde{\rho}_s = \sum_{\alpha, i} j_i^+(\alpha\omega_0)\left[\mathcal{G}_i^\alpha, \left[\mathcal{G}_i^{-\alpha}, \tilde{\rho}_s - \rho_0\right]\right]. \tag{4-117}$$

We call this approximation *intermediate* narrowing.

Finally, there is the so-called extreme narrowing limit where $c_\alpha^+(\tau)$ decays so rapidly that

$$e^{i\alpha\omega\tau} \simeq 1, \qquad \text{i.e.,} \quad \alpha\omega\tau_c \ll 1. \tag{4-118}$$

Then, $j^+(\alpha\omega_0)$ becomes

$$j^+(\alpha\omega_0) = j_\alpha^+(0) = \tfrac{1}{2}\int_{-\infty}^{+\infty} c_\alpha^+(\tau)\, d\tau \tag{4-119a}$$

and

$$-R\tilde{\rho}_s = \sum_{i, \alpha} j_i^+(0)\left[\mathcal{G}_i^\alpha, \left[\mathcal{G}_i^{-\alpha}, \tilde{\rho}_s - \rho_0\right]\right]. \tag{4-119b}$$

This is the level of the theory appropriate to nonviscous liquids relaxing via the random field interaction (4-99).

So far nothing has been said about the correlation functions which determine T_1 and T_t. The form which $c_\alpha^+(\tau)$ takes depends on what causes fluctuations in the local magnetic field. For example, these can arise from rotations, translations, or paramagnetic relaxation. A discussion of a number of correlation functions is given in Abragam's book.[11]

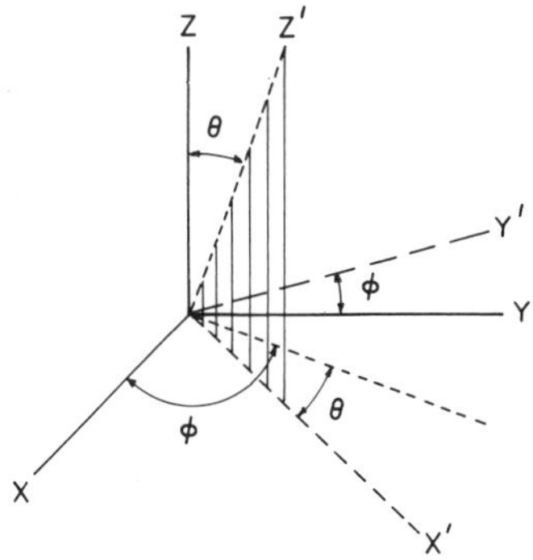

Fig. 4-1. The molecular coordinate system (x', y', z') in which uniaxial quadrupole and chemical shift interactions have axial symmetry. The molecular axis is taken as z'.

Relaxation mechanisms are characterized by their sets of $\mathcal{G}_i^\alpha$ (spin) and ν_i^α (lattice) operators together with their correlation functions. These matters are the subject of later sections of this chapter.

b. Quadrupole Relaxation[12, 13]

The interaction between nuclear quadrupole moments and fluctuating electric field gradients is the principal contribution to nuclear spin relaxation when $I > \frac{1}{2}$.

The quadrupole interaction of a uniaxial rotationally symmetric site in the coordinate system of the molecule (labeled by primes) is given (Fig. 4-1) as

$$\mathcal{H}_Q = A\left[3(I^{z'})^2 - I(I+1)\right], \tag{4-120}$$

where†

$$A = \frac{e^2qQ}{4I(2I-1)\hbar}, \tag{4-121}$$

$e^2qQ\hbar^{-1}$ is the quadrupole coupling constant and the electric field gradient is defined as

$$\delta^2V/\delta z'^2 = eq. \tag{4-122}$$

Note that for $I = \frac{1}{2}$, $\mathcal{H}_q$ is zero as then

$$(I_{z'}^z)^2 = \tfrac{1}{3}I(I+1). \tag{4-123}$$

We next write Eq. (4-124) in the laboratory frame using, see Fig. 4-1,

$$I_{z'}^{z'} = I^z\cos\theta + I^x\sin\theta\cos\varphi + I^y\sin\theta\sin\varphi. \tag{4-124}$$

†We assume axial symmetry, and we take the molecular axis to be z'.

The result is

$$\mathcal{H}_Q = A\{\tfrac{1}{2}(3\cos^2\theta - 1)(3(I^z)^2 - I(I+1))$$
$$+\tfrac{3}{2}\sin\theta\cos\theta(I^zI^+ + I^+I^z)e^{-i\varphi}$$
$$+\tfrac{3}{2}\sin\theta\cos\theta(I^zI^- + I^-I^z)e^{i\varphi}$$
$$+\tfrac{3}{4}\sin^2\theta e^{-i2\varphi}(I^+)^2 + \tfrac{3}{4}\sin^2\theta e^{i2\varphi}(I^-)^2\}, \tag{4-125}$$

which permits us to identify

$$\nu^0 = \tfrac{1}{2}(3\cos^2\theta - 1)A, \tag{4-126}$$
$$\nu^{\pm 1} = \tfrac{3}{2}\sin\theta\cos\theta e^{\mp i\varphi}A, \tag{4-127}$$
$$\nu^{\pm 2} = \tfrac{3}{4}\sin^2\theta e^{\mp i2\varphi}A, \tag{4-128}$$
$$\mathcal{G}^0 = 3I_z^2 - I(I+1), \tag{4-129}$$
$$\mathcal{G}^{\pm 1} = I^zI^{\pm} + I^{\pm}I^z, \tag{4-130}$$
$$\mathcal{G}^{\pm 2} = (I^{\pm})^2. \tag{4-131}$$

For a molecule which can be assumed to obey the isotropic rotational equation Abragam shows that the correlation function $c_\alpha^+(\tau)$ defined as

$$c_\alpha^+(\tau) = \mathrm{Tr_b}\, e^{-\beta \mathfrak{h}}\nu^\alpha(\tau)\nu^{-\alpha}(0) \tag{4-132}$$

is classically given as

$$c_\alpha^+(\tau) = \frac{1}{4\pi}\int_0^{2\pi}\int_0^{\pi} \nu^\alpha \nu^{-\alpha}\sin\theta\, d\theta\, d\varphi\, e^{-(\tau/\tau_q)}, \tag{4-133}$$

where τ_q is the quadrupole correlation time. Substituting into (4-133) for the ν_α's from (4-126) to (4-128) one finds that

$$c_0^+(\tau) = \tfrac{1}{5}e^{-(\tau/\tau_q)}A^2 = c_0^+(0)e^{-\tau/\tau_q}, \tag{4-134a}$$
$$c_{\pm 1}^+(\tau) = \tfrac{3}{10}e^{-(\tau/\tau_q)}A^2, \tag{4-134b}$$

and

$$c_{\pm 2}^+(\tau) = \tfrac{3}{10}e^{-(\tau/\tau_q)}A^2. \tag{4-134c}$$

In order to appreciate the different levels of approximation of R_q (quadrupole), it is enlightening to examine the form which an element of $R_q\rho_s$ takes. We start with R_q as it is expressed in (4-97). A typical element of R_q,

$$-\langle a|R_q\tilde{\rho}_s|b\rangle = \sum_{k,l} a_{k,l}(\tilde{\rho}_s - \rho_0)_{k,l}, \tag{4-135}$$

will consist of some linear combination of density matrix elements each weighted by the function $a_{k,l}$. These contain terms of the form

$$\frac{d}{2}\int_{-\infty}^{+\infty} c_{\alpha j}(0)e^{-(|\tau|/\tau_q)}e^{i\omega_j\alpha\tau}e^{i\Sigma_j b_j(\omega_{0j}-\omega)\tau}\,d\tau$$

$$= \frac{dc_\alpha^+(0)\tau_q}{1+\left[\alpha\omega+\Sigma_j b_j(\omega_{0j}-\omega)\right]^2\tau_q^2}, \tag{4-136}$$

where the b_j values come from elements of $e^{\pm i\mathcal{H}_s\tau}$ and the d's appear from elements of spin operators.

Intermediate narrowing involves dropping the $\omega_{0j}-\omega$ terms from the denominator on the right-hand side of (4-136) by assuming that

$$\sum_j b_j(\omega_{0j}-\omega) \ll \alpha\omega, \tag{4-137}$$

which is appropriate near resonance since $\omega_{0j}-\omega$ will be at most 10^4 rad sec^{-1}, while ω or ω_0 is between 10^7 and 7×10^{10} rad sec^{-1}. Dropping the $(\omega_{0j}-\omega)\tau_q$ terms (4-136) becomes

$$dc_\alpha^+(0)\tau_q/\left[1+(\alpha\omega_0)^2\tau_q^2\right]. \tag{4-138}$$

Finally, at the extreme narrowing limit we assume that

$$(\alpha\omega_0)^2\tau_q^2 \ll 1.$$

The relaxation operator at the intermediate narrowing approximation appears as (4-117)

$$R\tilde{\rho}_s = -\sum_{\alpha,i} j_{a,i}^+(\alpha\omega)\left[\mathcal{I}_i^\alpha,\left[\mathcal{I}_i^{-\alpha},\tilde{\rho}_s-\rho_0\right]\right], \tag{4-139}$$

where for one nucleus using (4-134a)–(4-134c),

$$j_{0,i}^+(0) = \tfrac{1}{5}A^2\tau_{qi} \tag{4-140}$$

$$j_{\pm1,i}^+(\omega_0) = \tfrac{3}{10}A^2\frac{1}{1+\omega^2\tau_{qi}^2}\tau_{qi} \tag{4-141}$$

$$j_{\pm2,i}^+(2\omega_0) = \tfrac{3}{10}A^2\frac{1}{1+4\omega^2\tau_{qi}^2}\tau_{qi}, \tag{4-142}$$

where A is given by (4-121). Thus, while each quadrupolar nucleus requires three spectral densities $j_\alpha^+(\alpha\omega)$, there are only two unknown parameters—the quadrupole coupling constant and the quadrupolar correlation time.

The definition of T_{1q} (if it exists) for a single spin is

$$-\frac{1}{T_{1q}}(\langle I^z\rangle-\langle I_0^z\rangle)$$

$$= \operatorname{Tr} I^z R_q\tilde{\rho}_s = \sum_\alpha j_\alpha^+(\alpha\omega)\operatorname{Tr}(\tilde{\rho}_s-\rho_0)\left[\mathcal{I}^\alpha,\left[\mathcal{I}^{-\alpha},I^z\right]\right]. \tag{4-143}$$

For $I = 1$ (4-143) is satisfied both for the intermediate and extreme narrowing limit, but for $I > 1$ (4-143) holds only in the extreme narrowing limit, the reason being that the double commutation relations needed in evaluating the right-hand side of (4-145),

$$[\mathcal{G}^1, [\mathcal{G}^{-1}, I^z]] = \{16(I^z)^3 - I^z[I(I+1) - 2]\}, \tag{4-144}$$

$$[\mathcal{G}^2, [\mathcal{G}^{-2}, I^z]] = \{-16(I^z)^3 + I^z[16I(I+1) - 8]\}, \tag{4-145}$$

contain an $(I^z)^3$ term which makes (4-143), in general, invalid. The exceptions will be for $I = 1$, where $(I^z)^3 = I^z$ and for extreme narrowing where the weighting functions

$$j_1^+(0) = j_2^+(0), \tag{4-146}$$

see (4-141) and (4-142), and thus the $(I^z)^3$ terms will cancel.

For $I = 1$ one obtains at intermediate narrowing

$$\frac{1}{T_{1q}} = 4j_1^+(\omega_0) + 16j_2^+(2\omega_0), \tag{4-147}$$

which at extreme narrowing becomes

$$\frac{1}{T_{1q}} = \frac{3}{8}\left(\frac{e^2qQ^2}{\hbar}\right)\tau_q \tag{4-148}$$

after substituting for A^2, see (4-121). In similar fashion, for $I = 1$, one obtains at intermediate narrowing

$$\frac{1}{T_{2q}} = 9j_0^+(0) + 10j_1^+(\omega) + 4j_2^+(2\omega), \tag{4-149}$$

and at extreme narrowing, i.e., $\omega_0\tau \to 0$, T_{2q} becomes equal to $T_{1q}(\omega_0\tau = 0)$. This equality of T_1 and T_2 is a general result of the extreme narrowing limit for all relaxation mechanisms, for an isolated spin.

c. Intramolecular Dipolar Relaxation (Intradipolar)[14, 15]

The required spin and lattice operators, for $\alpha = 0, \pm 1$, and ± 2, as given in (4-60) are

$$\mathcal{G}_{i,j}^0 = [I_i^z I_j^z - \tfrac{1}{4}(I_i^+ I_j^- + I_i^- I_j^+)], \tag{4-150}$$

$$\mathcal{G}_{i,j}^1 = I_i^+ I_j^- + I_i^- I_j^+ , \tag{4-151}$$

$$\mathcal{G}_{i,j}^2 = I_i^+ I_j^+ , \tag{4-152}$$

$$\nu_{i,j}^0 = k_{i,j}(1 - 3\cos^2\theta), \tag{4-153}$$

$$\nu_{i,j}^1 = -\tfrac{3}{2}k_{i,j}\sin\theta\cos\theta e^{-i\varphi}, \tag{4-154}$$

$$\nu_{i,j}^2 = -\tfrac{3}{4}k_{i,j}\sin^2\theta e^{-2i\varphi}, \tag{4-155}$$

where

$$k_{i,j} = \hbar^2 \gamma_i \gamma_j / r_{i,j}^3 \tag{4-156}$$

and $r_{i,j}$ is the distance between spins i and j.

The intradipolar relaxation operator has to be summed over all pairs of interacting spins in the molecule under consideration, i.e.,

$$R_d\tilde{\rho} = -\tfrac{1}{2} \sum_{\substack{\alpha(i,j)\\ \alpha(i',j')}}\sum \int_{-\infty}^{+\infty} c^{+}_{\substack{\alpha(i,j)\\ \alpha'(i',j')}} \left[\mathcal{G}^{\alpha}_{i,j}, e^{-i\mathcal{H}_s\tau}\left[\mathcal{G}^{-\alpha}_{i,j}, \tilde{\rho}_s - \tilde{\rho}_0 \right] e^{i\mathcal{H}_s} \right] d\tau. \tag{4-157}$$

The dominant interaction will be the autocorrelation ($i = i', j = j'$), the coupling of $v_{i,j}$ with $v_{i,j}$. For this case

$$c^{+}_{\alpha\,(i,j)}(\tau) = \mathrm{Tr}_b e^{-\beta\mathcal{H}_b} v^{\alpha}_{i,j}(\tau) v^{-\alpha}_{i,j}(0). \tag{4-158a}$$

As is the case for quadrupolar relaxation, the rotational classical autocorrelation function is given as

$$c^{+}_{\alpha\,(i,j)}(\tau) = \frac{1}{4\pi} \int_0^{2\pi} \int_0^{\pi} v^{\alpha}_{i,j}(0) v^{-\alpha}_{i,j}(0) \sin\theta \, d\theta \, d\phi \, e^{-(\tau/\tau_d)}, \tag{4-158b}$$

where τ_d is the rotational intradipolar correlation time. Using (4-153) to obtain (4-156) and (4-158b) the three correlation functions are

$$c^{+}_{0\,(i,j)}(\tau) = \tfrac{4}{5} k^2_{i,j} e^{-(\tau/\tau_d)}, \tag{4-159}$$

$$c^{+}_{1\,(i,j)}(\tau) = \tfrac{3}{5} k^2_{i,j} e^{-(\tau/\tau_d)}, \tag{4-160}$$

$$c^{+}_{2\,(i,j)}(\tau) = \tfrac{3}{10} k^2_{i,j} e^{-(\tau/\tau_d)}. \tag{4-161}$$

The spectral densities $j^{+}_{i,j}(\alpha\omega)$ at the level of intermediate narrowing are

$$j^{+}_{i,j}(\alpha_i\omega_i) = \frac{c^{+}_{\alpha\,(i,j)}(0)\tau_d}{1 + (\alpha_i\omega_i)^2\tau_d^2}, \qquad \alpha = 0, \pm 1, \pm 2. \tag{4-162}$$

Each internuclear interaction is characterized by three spectral densities which are functions of the gyromagnetic ratios, the internuclear distances, and the τ_ds. Where reasonable assumptions may be made about structure, the only variable is τ_d.

Finally, it must be mentioned that intradipolar relaxation is responsible for the intramolecular Overhauser effect observed in double resonance experiments (to be discussed in Chapter VIII).

d. Intermolecular Dipole Relaxation[16–21]

Intermolecular dipole relaxation contributes to the random field relaxation operator (4-99) and gives rise to the intermolecular Overhauser effect seen in double resonance.

Consider a collection of molecules of two kinds, A and B. They collide with one another and their close interactions contribute to relaxation. To describe this effect we start with the density matrix equation for all spins in all A and B molecules given as

$$\dot{\tilde{\rho}}^{AB} = -i\left[\overline{\mathcal{H}}^{A} + \overline{\mathcal{H}}^{B}, \tilde{\rho}^{AB}\right] + R_{AB}\tilde{\rho}^{AB} + R_{AA}\tilde{\rho}^{AB} + R_{BB}\tilde{\rho}^{AB}, \tag{4-163}$$

where the intermolecular relaxation operator between molecules A and B is

$$R_{AB}\tilde{\rho}^{AB} = -\tfrac{1}{2}\sum_{n,m}\sum_{\substack{(i,j)\\(i',j')}}\int_{-\infty}^{+\infty} d\tau\, e^{i\omega\alpha\tau} C^{+n,m}_{\substack{(i,j)\\(i',j')}}(\tau)$$

$$\times\left[\mathcal{G}^{\alpha}_{i,j}, e^{-i\overline{\mathcal{H}}^{AB}\tau}\left[\mathcal{G}^{-\alpha}_{i',j'}, \tilde{\rho}^{AB} - \tilde{\rho}^{AB}_0\right]e^{i\overline{\mathcal{H}}^{AB}\tau}, \tag{4-164}$$

where n sums over A molecules, m over B molecules, i, i' are spins in an A molecule, and j, j' in a B molecule. The ν's and $\mathcal{G}$'s for intermolecular dipolar interaction are

$$\mathcal{G}^{0}_{i,j} + \left[I_i^z I_j^z - \tfrac{1}{4}(I_i^+ I_j^- + I_i^+ I_j^-)\right], \tag{4-165}$$

$$\nu^{0}_{i,j} = \left(\mu^2/r^2_{i,j}\right)(1 - 3\cos^2\theta), \tag{4-166}$$

$$\mathcal{G}^{+1}_{i,j} = \left[I_i^+ I_j^z + I_i^z I_j^+\right], \tag{4-167}$$

$$\nu^{+1}_{i,j} = -\left(3\mu^2/2r^3_{i,j}\right)(\sin\theta\cos\theta\, e^{-i\varphi}), \tag{4-168}$$

$$\mathcal{G}^{-1}_{i,j} = \left(\mathcal{G}^{+1}_{i,j}\right)^{\dagger}, \tag{4-169}$$

$$\nu^{-1}_{i,j} = \left(\nu^{+1}_{i,j}\right)^{*}, \tag{4-170}$$

$$\mathcal{G}^{+2}_{i,j} = I_i^+ I_j^-, \tag{4-171}$$

$$\nu^{+2}_{i,j} = -\left(3\mu^2/4r^3_{i,j}\right)(\sin^2\theta e^{-2i\varphi}), \tag{4-172}$$

$$\mu^2 = \hbar^2\gamma_i\gamma_j. \tag{4-173}$$

Next, by analogy with (4-21) we write

$$\tilde{\rho}^{AB} = \tilde{\rho}^{A}\tilde{\rho}^{B} + g, \tag{4-174}$$

$$\mathrm{Tr}_A\, g = \mathrm{Tr}_B\, g = 0. \tag{4-175}$$

Substituting Eq. (4-174) into Eq. (4-163) and taking the spin trace over B one obtains $\dot{\rho}$ for all A molecules,

$$\dot{\tilde{\rho}}^{A} = -i\left[\overline{\mathcal{H}}^{A}, \tilde{\rho}^{A}\right] + \mathrm{Tr}_B\, R_{AB}\tilde{\rho}^{AB} + R_{AA}\,\mathrm{Tr}_b\,\tilde{\rho}^{AB} \tag{4-176}$$

$$\simeq -i\left[\overline{\mathcal{H}}^{A}, \tilde{\rho}^{A}\right] + \mathrm{Tr}_B\, R_{AB}\tilde{\rho}^{A}\tilde{\rho}^{B} + R_{AA}\tilde{\rho}^{A}. \tag{4-176a}$$

The neglected g from (4-174) would lead to higher-order couplings in the second term.

What we want is the density matrix equation for a single molecule. This is obtained from Eq. (4-182) by operating on it with $\mathrm{Tr}'_{\mathrm{A}}$, where the prime indicates the trace is to be taken over all A molecules but one (labeled n); the result is

$$\dot{\tilde{\rho}}^{\mathrm{A}_n} = -i\left[\overline{\mathcal{H}}^{\mathrm{A}_n}, \tilde{\rho}^{\mathrm{A}_n}\right] + R_{\mathrm{A}_n}\tilde{\rho}^{\mathrm{A}_n} + N_{\mathrm{B}}\,\mathrm{Tr}_{\mathrm{B}_m}\, R_{\mathrm{A}_n\mathrm{B}_m}\tilde{\rho}^{\mathrm{A}_n}\tilde{\rho}^{\mathrm{B}_m}, \quad (4\text{-}177)$$

where N_{B} is the total number of B molecules in the sample, n refers to just one of the A molecules, and m is a B molecule. Noting that as $R_{\mathrm{A}_n\mathrm{B}_m}$ is defined (4-70) as a trace over *both* coordinates of A and B, we can more meaningfully rewrite (4-177) as

$$\tilde{\rho}^{\mathrm{A}_n} = -i\left[\overline{\mathcal{H}}^{\mathrm{A}_n}, \tilde{\rho}^{\mathrm{A}_n}\right] + R_{\mathrm{A}_{nn}}\tilde{\rho}^{\mathrm{A}_n} + (\mathrm{B})\mathrm{Tr}_{\mathrm{B}_m}\, \overline{R}_{\mathrm{A}_n\mathrm{B}_m}\tilde{\rho}^{\mathrm{A}_n}\tilde{\rho}^{\mathrm{B}_m} \quad (4\text{-}178)$$

where the bar on $\overline{R}_{\mathrm{A}_n\mathrm{B}_m}$ indicates an integration over the relative coordinates of molecules A_n and B_m. The dominant contribution to the intermolecular relaxation will be from the autocorelation terms for which

$$\overline{R}_{\mathrm{A}_n\mathrm{B}_m} = \sum_{i,j} R_{i,j}\,\tilde{\rho}^{\mathrm{A}_n}\tilde{\rho}^{\mathrm{B}_m} \quad (4\text{-}179)$$

$$R_{i,j}\,\tilde{\rho}^{\mathrm{A}_n\mathrm{B}_m} = -\frac{1}{2}\sum_{\alpha}\int_{-\infty}^{\infty} d\tau\, e^{i\omega\alpha\tau} c_{\alpha}^{+}(r_{i,j}, t)$$

$$x\left[\mathcal{G}_{i,j}^{\alpha}, e^{-i(\overline{\mathcal{H}}^{\mathrm{A}_n}+\overline{\mathcal{H}}^{\mathrm{B}_n})\tau}\left[\mathcal{G}_{i,j}^{-\alpha}, \tilde{\rho}^{\mathrm{A}_n}\tilde{\rho}^{\mathrm{B}_m} - \tilde{\rho}_0^{\mathrm{A}_n}\tilde{\rho}_0^{\mathrm{B}_m}\right] e^{i(\overline{\mathcal{H}}^{\mathrm{A}_n}+\overline{\mathcal{H}}^{\mathrm{B}_u})\tau}\right]. \quad (4\text{-}180)$$

Equation (4-180), as it appears, is bilinear. We linearize it by writing

$$\tilde{\rho}^{\mathrm{A}_n} = \rho_0^{\mathrm{A}_n} + \rho_1^{\mathrm{A}_n} \quad (4\text{-}181)$$

thus

$$\tilde{\rho}^{\mathrm{A}_n}\tilde{\rho}^{\mathrm{B}_m} - \rho_0^{\mathrm{A}_n}\rho_0^{\mathrm{B}_m} = \rho_1^{\mathrm{A}_n}\rho_0^{\mathrm{B}_m} + \rho_1^{\mathrm{B}_m}\rho_0^{\mathrm{A}_n}, \quad (4\text{-}182)$$

neglecting the products $\rho_1^{\mathrm{A}_n}\rho_1^{\mathrm{B}_m}$ where they appear in (4-178).

In general,

$$J\tau_c \ll 1, \qquad \gamma\mathcal{H}_{\mathrm{rf}}\tau_c \ll 1, \quad (4\text{-}183)$$

where τ_c is the correlation time of the relaxation process. We can thus replace $\overline{\mathcal{H}}^{\mathrm{A}_n}$ and $\overline{\mathcal{H}}^{\mathrm{B}_n}$ appearing in the exponentials of Eq. (4-180) by

$$\overline{\mathcal{H}}^{\mathrm{A}_n} = \sum_i (\omega_{0i} - \omega) I_i^z, \quad (4\text{-}184)$$

$$\overline{\mathcal{H}}^{\mathrm{B}_m} = \sum_j (\omega_{0j} - \omega) I_j^z. \quad (4\text{-}185)$$

To make the form of the operator in (4-177) apparent we write (4-181) in the intermediate narrowing approximation where

$$\Delta\omega\,\tau \ll 1, \quad (4\text{-}186)$$

and use (4-182) . The result is

$$R_{i,j}\tilde{\rho}^{A_n}\tilde{\rho}^{B_m} = -\tfrac{1}{2}\sum_{\alpha}\int_{-\infty}^{+\infty} e^{i\omega\alpha\tau}c_{\alpha}^{+}(\mathbf{r}_{i,j},\tau)$$

$$\times\left[\mathcal{G}_{i,j}^{\alpha},\left[\mathcal{G}_{i,j}^{-\alpha},\rho_1^{A_n}\rho_0^{B_m}+\rho_1^{B_m}\rho_0^{A_n}\right]\right]d\tau. \quad (4\text{-}187)$$

Each $\mathcal{G}_{i,j}^{\alpha}$ can be written as [see (4-165), (4-176), (4-171)]

$$\mathcal{G}_{i,j}^{\alpha} = \sum_{n,m}\mathcal{G}_i^n(\alpha)\mathcal{G}_j^m(\alpha). \quad (4\text{-}188)$$

To be specific, consider

$$\mathcal{G}_{i,j}^{2} = (I_i^{+})^{A_n}(I_j^{+})^{B_m}. \quad (4\text{-}189)$$

In (4-187) we thus will have one term of the form

$$-\tfrac{1}{2}\int_{-\infty}^{+\infty} e^{i\omega\alpha\tau}c_2^{+}(\mathbf{r}_{i,j},\tau)\left[I_i^{+}I_j^{+},\left[I_i^{-}I_j^{-},\rho_1^{A_n}\rho_0^{B_m}+\rho_1^{B_m}\rho_0^{A_n}\right]\right]d\tau \quad (4\text{-}190)$$

which using the high temperature approximation

$$\rho_0^{A(B)} \simeq 1/N_{A(B)} \quad (4\text{-}191)$$

becomes upon substituting into (4-178)

$$-\tfrac{1}{2}(B)\int_{-\infty}^{+\infty} e^{i\omega\alpha\tau}c_2^{+}(\mathbf{r}_{i,j},\tau)$$

$$\times\left\{\frac{1}{N_B}\left[I_i^{+},\left[I_i^{-},\rho_1^{A_n}\right]\right]\mathrm{Tr}_{B_m} I_j^{+}I_j^{-} + \frac{1}{N_A}2I_i^z\,\mathrm{Tr}_{B_m}\rho_1^{B_m}2I_j^z\right\}d\tau. \quad (4\text{-}192)$$

The other terms are derived in a similar fashion. Note that the first term in braces appears as a random field relaxationlike expression (4-6) and that the second term gives no contribution in the low power limit.

e. Relaxation via Anisotropy of the Chemical Shift

In the coordinate system of the spin symmetry axis of a particular nucleus (denoted by primes) the uniaxial Zeeman (Ze) interaction is

$$\mathcal{H}_{Ze} = \gamma_l H_{z'}I^{z'} + \gamma_t H_{x'}I^{x'} + \gamma_t H_{y'}I^{y'}$$

$$= \mu g_l H_{z'}I^{z'} + \mu g_t H_{x'}I^{x'} + \mu g_t H_{y'}I^{y'}, \quad (4\text{-}193)$$

where l means longitudinal and t means transverse. In the coordinate system of the spin symmetry axes the magnetic field components are given as, Fig. 4-1,

$$\mathcal{H}_{z'} = (\cos\theta)H_0, \qquad \mathcal{H}_{x'} = -(\sin\theta)H_0, \qquad \mathcal{H}_{y'} = 0, \quad (4\text{-}194)$$

and one has that

$$\mathcal{H}_{\mathrm{Ze}} = \omega_{0l} I^{z'} \cos\theta - \omega_{0t} I^{x'} \sin\theta. \tag{4-195}$$

We next go back to the laboratory frame by noting that

$$I^{z'} = I^z \cos\theta + I^x \sin\theta\cos\varphi + I^y \sin\theta \sin\varphi, \tag{4-196}$$

$$I^{x'} = -I^z \sin\theta + I^x \cos\theta\cos\varphi + I^y \cos\theta\sin\varphi. \tag{4-197}$$

Using

$$I^x = \tfrac{1}{2}(I^+ + I^-), \qquad I^y = -\tfrac{1}{2} i (I^+ - I^-), \tag{4-198}$$

and combining (4-193)–(4-197) one obtains

$$\begin{aligned}\mathcal{H}_{\mathrm{Ze\,lab}} = \omega_{0l} \cos\theta\Big[I^z \cos\theta + \tfrac{1}{2} I^+ \sin\theta(\cos\varphi - i\sin\varphi) \\ + \tfrac{1}{2} I^- \sin\theta(\cos\varphi + i\sin\varphi)\Big] - \omega_{0t}\sin\theta\Big[-I^z\sin\theta + \tfrac{1}{2}I^+\cos\theta \\ \times(\cos\varphi - i\sin\varphi) + \tfrac{1}{2}I^-\cos\theta(\cos\varphi + i\sin\varphi)\Big]\end{aligned} \tag{4-199}$$

or

$$\begin{aligned}\mathcal{H}_{\mathrm{Ze\,lab}} &= \big[\omega_{0l}\cos^2\theta + \omega_{0t}\sin^2\theta\big] I^z \\ &\quad + \sin\theta\cos\theta(\omega_{0l} - \omega_{0t})\tfrac{1}{2} e^{-i\varphi} I^+ \\ &\quad + \sin\theta\cos\theta(\omega_{0l} - \omega_{0t})\tfrac{1}{2} e^{-i\varphi} I^-.\end{aligned} \tag{4-200}$$

We next write (z laboratory = zL)

$$\mathcal{H}_{\mathrm{zL}} = \langle \mathcal{H}_{\mathrm{zL}} \rangle + \mathcal{H}_{\mathrm{zL}} - \langle \mathcal{H}_{\mathrm{zL}} \rangle = \langle \mathcal{H}_{\mathrm{zL}} \rangle + \mathcal{H}_{\mathrm{spin\ lat}}, \tag{4-201}$$

where

$$\mathcal{H}_{\mathrm{spin\ lat}} = \mathcal{H}_{\mathrm{zL}} - \langle \mathcal{H}_{\mathrm{zL}} \rangle \tag{4-202}$$

and

$$\langle \cdots \rangle = \frac{1}{4\pi}\int \cdots \sin\theta\, d\theta\, d\varphi \tag{4-203}$$

means angular average. Thus, we have

$$\begin{aligned}\langle \mathcal{H}_{\mathrm{zL}} \rangle &= \frac{1}{4\pi}\int \big[\omega_{0l}\cos^2\theta + \omega_{0t}(1 - \cos^2\theta)\big]\sin\theta\, d\theta\, d\varphi\, I^z \\ &= \tfrac{1}{2}\int_{-1}^{+1}\big[\omega_{0l}x^2 + \omega_{0t}(1 - x^2)\big]\, dx\, I^z \\ &= \tfrac{1}{3}(\omega_{0l} + 2\omega_{0t}) I^z\end{aligned} \tag{4-204}$$

and $\mathcal{H}_{\mathrm{spin\ lat}}$,

$$\begin{aligned}\mathcal{H}_{\mathrm{spin\ lat}} &= \big[\omega_{0l}\big(\cos^2\theta - \tfrac{1}{3}\big) + \omega_{0t}\big(\sin^2\theta - \tfrac{2}{3}\big)\big] I^z \\ &\quad + (\omega_{0l} - \omega_{0t})\sin\theta\cos\theta\tfrac{1}{2}e^{-i\varphi}I^+ + (\omega_{0l} - \omega_{0t})\sin\theta\cos\theta\tfrac{1}{2}e^{i\varphi}I^- \\ &= \sum_\alpha \nu^\alpha \mathcal{G}^\alpha,\end{aligned} \tag{4-205}$$

where

$$\nu^0 = \left[\omega_{01}\left(\cos^2\theta - \tfrac{1}{3}\right) + \omega_{0t}\left(\sin^2\theta - \tfrac{2}{3}\right)\right], \qquad \mathcal{I}^0 = I^z, \quad (4\text{-}206)$$

$$\nu^{\pm 1} = (\omega_{01} - \omega_{0t})\sin\theta\cos\theta\tfrac{1}{2}e^{\mp i\varphi}, \quad (4\text{-}207)$$

$$\mathcal{I}^{\pm 1} = I^{\pm}. \quad (4\text{-}208)$$

3. Recapitulation of Section 1 and Summary

The contribution of nuclear spin relaxation to the density matrix equation

$$\dot{\tilde{\rho}}_s = -i[\mathcal{H}, \tilde{\rho}_s] + \sum_m R_m\tilde{\rho}_s \quad (4\text{-}209)$$

is contained in the relaxation operators R_m, where m sums over the different relaxation mechanisms.

The interaction between the nuclear spin and its environment, the bath, is given by the Hamiltonian

$$\mathcal{H}_{sb} = \sum_{\alpha, i}^{k} \mathcal{I}_i^\alpha \nu_i^\alpha, \quad (4\text{-}210)$$

where $\mathcal{I}_i^\alpha$ and ν_i^α are spin and bath operators, respectively, k is the rank, and α the order of the spherical tensor, α runs from $k, k-1, \ldots,$ to $-k$, i sums over spins and sb means spin bath. Since $\mathcal{H}_{sb}$ is hermitian, the $\mathcal{I}$ and ν operators have the properties

$$\mathcal{I}_i^{-\alpha} = (\mathcal{I}_i^\alpha)^\dagger \quad (4\text{-}211)$$

and

$$\nu_i^{-\alpha} = (\nu_i^\alpha)^*. \quad (4\text{-}212)$$

The general expression for the relaxation operator is

$$R\tilde{\rho}_s = -\tfrac{1}{2}\sum_{\alpha(i,i')}\sum \int_{-\infty}^{+\infty} d\tau\, e^{i\omega\alpha\tau}$$

$$\times\left\{c^{+}_{\alpha(i,i')}(\tau)\left[\mathcal{I}_i^\alpha, e^{-i\mathcal{H}_s\tau}\left[\mathcal{I}_i^{-\alpha}, \tilde{\rho}_s\right]e^{i\mathcal{H}_s\tau}\right]\right.$$

$$\left. + c^{-}_{\alpha(i,i')}(\tau)\left[\mathcal{I}_i^\alpha, e^{-i\mathcal{H}_s\tau}\left[\mathcal{I}_i^{-\alpha}, \tilde{\rho}_s\right]_+ e^{i\mathcal{H}_s\tau}\right]\right\}, \quad (4\text{-}213)$$

where the correlation functions are

$$c^{\pm}_{\alpha(i,i')}(\tau) = \tfrac{1}{2}\left[c_{\alpha(i,i')}(+\tau) \pm c_{-\alpha(i',i)}(-\tau)\right], \quad (4\text{-}214)$$

$$c_{\alpha(i,i')}(\tau) = \mathrm{Tr}_b\, \rho_0(\mathcal{H}_b)\nu_i^\alpha(\tau)\nu_{i'}^{-\alpha}, \quad (4\text{-}215)$$

and

$$\nu_i^\alpha(\tau) = e^{i\mathcal{H}_b\tau}\nu_i^\alpha e^{-i\mathcal{H}_b\tau}. \tag{4-216}$$

It is generally acceptable to approximate $\mathcal{H}_s$ in (4-213) as

$$\overline{\mathcal{H}}_s \simeq \sum_i (\omega_{0i} - \omega) I_i^z. \tag{4-217}$$

The operator described in (4-213) can be handled at several levels of approximation. It is shown that for $\hbar\beta\omega_0 \ll 1$ one can express (4-213) as

$$R\tilde{\rho}_s = -\tfrac{1}{2}\sum_{\alpha(i,i')}\sum\left\{\left(\int_{-\infty}^{+\infty} d\tau\, c_{\alpha(i,i')}^{+}(\tau) e^{i\omega_\alpha\tau}\left[\mathcal{G}_i^\alpha, e^{-i\overline{\mathcal{H}}_s\tau}\left[\mathcal{G}_{i'}^{-\alpha}, \tilde{\rho}_s - \rho_0\right] e^{i\overline{\mathcal{H}}_s\tau}\right]\right\}, \tag{4-218}$$

where ρ_0 is the equilibrium density matrix.

Near resonance the exponential operators in (4-218) can be set equal to unity assuming $\Delta\omega\tau_{\text{correlation}} \ll 1$, simplifying (4-222) as

$$R\tilde{\rho}_s = \tfrac{1}{2}\sum_{\alpha(i,i')}\sum \int_{-\infty}^{+\infty} c_{\alpha(i,i')}^{+}(\tau) e^{i\omega_\alpha\tau}\left[\mathcal{G}_i^\alpha, \left[\mathcal{G}_{i'}^{-\alpha}, \tilde{\rho}_s - \rho_0\right]\right]. \tag{4-219}$$

Finally, the extreme narrowing approximation

$$\omega\tau_c \ll 1 \tag{4-220}$$

allows (4-219) to appear as

$$R\tilde{\rho}_s = -\tfrac{1}{2}\sum_{\alpha(i,i')}\sum \int_{-\infty}^{+\infty} c_{\alpha(i,i')}^{+}(\tau)\, d\tau\left[\mathcal{G}_i^\alpha, \left[\mathcal{G}_{i'}^{-\alpha}, \tilde{\rho}_s - \rho_0\right]\right]. \tag{4-221}$$

Relaxation due to intramolecular dipole-dipole, intermolecular dipole-dipole, electric quadrupolar, and chemical shift anisotropy interactions, respectively, are discussed and appropriate spin and lattice operators defined.

Appendix A. A Theorem about the Trace

To prove

$$\begin{aligned}
\mathrm{Tr}_b[A(b), B(b,s)] &= 0 \\
&= \mathrm{Tr}_b(A(b)B(b,s) - B(b,s)A(b)) \\
&= \sum_{b,b'} \{\langle b|A(b)|b'\rangle\langle b'|B(b,s)b\rangle \\
&\quad -\langle b|(b,s)|b'\rangle\langle b'|A)b)|b\rangle\}.
\end{aligned} \tag{4-A1}$$

As b and b' are dummy indices we can change $b \to b'$ and $b' \to b$ where they occur in the second summation. Thus, the right-hand side of (4-A1) becomes

$$\sum_{b,b'} \{\langle b|A(b)|b'\rangle\langle b'|B(b,s)|b\rangle - \langle b'|B(b,s)|b\rangle\langle b|A(b)|b'\rangle\}. \tag{4-A2}$$

Since $\langle b|\mathrm{A}(b)|b\rangle$ is a constant in s space, we can reverse the order of the matrix elements in the second term of (4-A2), giving

$$\sum_{b,\, b'} \{\langle b|A(b)b'\rangle\langle b'|B(b, s)b\rangle - \langle b|A(b)b'\rangle\langle b'|B(b, s)|b\rangle\},$$

which is identically equal to zero.

Appendix B. Glossary of Symbols

$c_\alpha^\pm(\tau)$	Correlation function
i, j, s, t	Spins
$j^\pm(\alpha\omega)$	Spectral density
k	Boltzmann constant
t	Time
I^x, I^y, I^z, I^+, I^-	Spin operators
N	Number of states
R	Relaxation operator
rf	Radio frequency
T, T_b	Temperature
T_1, T_l	Longitudinal relaxation time
T_t, T_2	Transverse relaxation time
$\mathcal{H}$	Hamiltonian, total
$\mathcal{H}_b$	Hamiltonian, bath
$\mathcal{H}_{sb}$	Spin bath interaction
$\bar{\mathcal{H}}_{rf}$	Driving Hamiltonian, rotating frame
$\bar{\mathcal{H}}_s$	Spin Hamiltonian, rotating frame
$\mathcal{I}_i^\alpha$	Spin operator
$\mathcal{L}$	Liouville operator—Subscript: s (spin), rf (driving term), b (bath), sb (spin bath)
$\bar{\mathcal{L}}$	Liouville operator, rotating frame
$\mathcal{P}$	Projection operator
α	Order of tensor
β	$(RT)^{-1}$
η	Derivation operator
θ, φ	Spherical coordinates
ν^α	Bath operator
ρ_T	Total density matrix
ρ_s	Spin density matrix, also called ρ for simplicity
$\tilde{\rho}_s$	Spin density matrix, rotating frame
τ_c	Correlation time
τ_q	Quadrupole correlation time
ω	Frequency, rad sec^{-1}
ω_1	rf power

Problems

1. Show that the Bloch equation form for relaxation (1-8) and (1-9) follows from (4-6).
2. Verify Eq. (4-7).
3. Verify Eqs. (4-43).
4. Go from Eq. (4-70) to Eq. (4-72).
5. Go from Eq. (4-72) to Eq. (4-74).
6. Go from Eq. (4-74) to Eq. (4-78).
7. Obtain $\mathcal{H}_{sb}$ from the interaction
$$\mathcal{H}_s = J_0 I_i^z I_j^z + J_i(I_i^x I_j^x + I_i^y I_j^y)$$
given in the molecular coordinate system.
8. Obtain the density matrix equations needed to calculate the NMR lineshape for a single spin, $I = \frac{3}{2}$, for example ^{7}Li, low rf field, subject to quadrupole relaxation, extreme narrowing approximation.
9. Verify Eq. (4-136).
10. Evaluate all elements of $R_d\tilde{\rho}$, dipole–dipole relaxation, for a two spin $I = \frac{1}{2}$, homonuclear system AB.

REFERENCES

1. P. N. Argyres and P. L. Kelly, *Phys. Rev.* **34**, A98 (1964).
2. F. Bloch, *Phys. Rev.* **102**, 104 (1956).
3. F. Bloch, *Phys. Rev.* **105**, 1206 (1957).
4. J. I. Kaplan, *J. Chem. Phys.* **55**, 1489 (1971).
5. F. Noack, in "NMR Basis Principles are Progress" (P. Diehl, E. Fluck, and R. Kosfeld, eds.), Vol. 3, p. 82. Springer-Verlag, Berlin and New York, 1971.
6. Y. Ayant, *J. Phys. Radium* **16**, 411 (1955).
7. J. F. Harman and B. H. Miller, *Phys. Rev.* **182**, 400 (1969).
8. P. S. Hubbard, *Phys. Rev.* **131**, 1155 (1963).
9. N. C. Pyper, *Mol. Phys.* **21**, 1 (1971).
10. N. C. Pyper, *Mol. Phys.* **21**, 961 (1971).
11. A. Abragam, "The Principles of Nuclear Magnetism," Chap. 8. Oxford Univ. Press, London and New York, 1961.
12. N. C. Pyper, *Mol. Phys.* **21**, 977 (1971).
13. N. C. Pyper, *Mol. Phys.* **21**, 1 (1971).
14. N. C. Pyper, *Mol. Phys.* **22**, 433 (1971).
15. P. S. Hubbard, *Rev. Mod. Phys.* **33**, 249 (1961).
16. I. Solomon, *Phys. Rev.* **99**, 559 (1955).
17. T. N. Khazanovich and V. Y. Zitserman, *Mol. Phys.* **21**, 65 (1971).
18. M. Alla and E. Lippma, *J. Magn. Reson.* **4**, 241 (1971).
19. N. R. Krishna and S. L. Gordon, *J. Chem. Phys.* **58**, 5687 (1973).

20. N. R. Krishna and S. L. Gordon, *J. Chem. Phys.* **59**, 4569 (1973).
21. A. Abragam and R. V. Pound, *Phys. Rev.* **92**, 943 (1953).

GENERAL REFERENCES

A. G. Redfield, *IBM Res. Dev.* **1**, 19 (1957); H. G. Redfield, *Adv. Magn. Reson.* **1**, 1 (1965). Redfield relaxation theory.

A. Abragam, "The Principles of Nuclear Magnetism," Chap. VIII. Oxford Univ. Press, London and New York, 1961.

C. P. Slichter, "Principles of Magnetic Resonance," Chaps. 5 and 6. Harper, New York, 1964.

R. K. Wangness and F. Bloch, *Phys. Rev.* **89**, 728 (1953).

R. Zwanzig, *J. Chem. Phys.* **33**, 1338 (1960). Irreversible process theory.

U. Fano, *J. Chem. Phys.* **33**, 1338 (1960). Transition probabilities, pressure broadening, Zwanzig operator.

N. Bloembergen, E. M. Purcell, and R. V. Pound, *Phys. Rev.* **73**, 679 (1948).

M. C. Wang and G. E. Uhlenbeck, *Rev. Mod. Phys.* **17**, 323 (1945). Theory of Brownian motion, hermitian herturbation, random in time.

I. Oppenheim and M. Bloom, *Can. J. Chem.* **39**, 895 (1961). Correlation function, hard sphere.

J. Korringa, J. L. Motchane, P. Papon, and Y. Yoshimori, *Phys. Rev.* **133**, 1230 (1964). Time dependent Hamiltonian, modified Bloch equations.

J. H. Freed and G. K. Fraenkel, *J. Chem. Phys.* **39**, 326 (1963). Relaxation matrix, organic radicals.

R. M. Lynden-Bell, *Mol. Phys.* **22**, 837 (1971). Slow orientation.

R. A. Hoffmann, *Advan. Magn. Reson.* **4**, 87 (1970).

M. Leigh, *J. Mag. Reson.* **4**, 308 (1971).

L. G. Werbelow and D. M. Grant, *Advan. Magn. Reson.* **9**, 190 (1977).

Chapter V
CHEMICAL REORGANIZATION

1. The Effects of Rate Processes[1-3]

In the previous chapter it was shown that the nuclear spin system of a molecule which interacts with a random magnetic field obeys the density matrix equation in the rotating frame

$$\dot{\tilde{\rho}} = -i\left[\overline{\mathcal{H}}, \tilde{\rho}\right] + R(\tilde{\rho} - \tilde{\rho}_0), \tag{5-1}$$

where the relaxation operator is given by

$$\begin{aligned} -R(\tilde{\rho} - \tilde{\rho}_0) = \sum_s \Bigg\{ & \frac{1}{T_{\mathrm{ls}}}\left[I_s^+, \left[I_s^-, \tilde{\rho} - \tilde{\rho}_0\right]\right] \\ & + \frac{1}{T_{\mathrm{ls}}}\left[I_s^-, \left[I_s^+, \tilde{\rho} - \tilde{\rho}_0\right]\right] \\ & + \frac{1}{T_{\mathrm{ts}}}\left[I_s^z, \left[I_s^z, \tilde{\rho} - \tilde{\rho}_0\right]\right] \Bigg\}, \end{aligned} \tag{5-2}$$

where s sums over spins, l means longitudinal, and t transverse, and

$$\tilde{\rho}_0 = \frac{e^{-\mathcal{H}/kT}}{\mathrm{Tr}\, e^{-\mathcal{H}/kT}}. \tag{5-3}$$

We shall now discuss the form of the density matrix equation which applies to an equilibrium system undergoing some kind of chemical reorganization. This blanket term covers rotation, pseudorotation, Berry rotation, Walden inversion, and all other intramolecular rearrangments, as well as bimolecular exchange.

Consider an equilibrium exchanging system where two species, AB and CD, mutually exchange two different fragments, B with C, in a bimolecular collision. For reasons which become apparent later it is convenient to describe the molecules in terms of the fragments which exchange

$$\mathrm{AB} + \mathrm{CD} \underset{k_r}{\overset{k_f}{\rightleftarrows}} \mathrm{AC} + \mathrm{BD}. \tag{5-4}$$

The fragments A, B, C, and D can be single protons or more complex moieties. Throughout this presentation it is important to be aware that we are always dealing with equilibrium systems.

The mean lifetime between exchanges τ_{sp} of each species involved in the equilibrium process is defined as the ratio of its rate of formation R_{ex} (or disappearance, they are the same) to its concentration[4] (sp).

$$1/\tau_{sp} = R_{ex}/(\mathrm{sp}). \tag{5-5}$$

Then the τs for the different species are given

$$\begin{aligned} 1/\tau_{AB} &= k_f(\mathrm{CD}), \qquad 1/\tau_{CD} = k_f(\mathrm{AB}), \\ 1/\tau_{AC} &= k_r(\mathrm{BD}), \qquad 1/\tau_{BD} = k_r(\mathrm{AC}), \\ k_f/k_r &= (\mathrm{AC})(\mathrm{BD})/(\mathrm{AB})(\mathrm{CD}). \end{aligned} \tag{5-6}$$

We can pictorially rewrite reaction (5-4) as

$$\mathrm{AB} + \mathrm{CD} \rightleftharpoons \underset{\mathbf{1}}{\left(\frac{\mathrm{AB}}{\mathrm{CD}}\right)} \rightleftharpoons \underset{\mathbf{2}}{\left(\begin{array}{c|c} \mathrm{A} & \mathrm{B} \\ \mathrm{C} & \mathrm{D} \end{array}\right)} \rightleftharpoons \mathrm{AC} + \mathrm{BD} \tag{5-7}$$

to proceed via two intermediate bimolecular complexes, **1**, before new bonds are formed and **2** after rearrangement but before dissociation to products.

Let us concentrate on AB for the moment. The nuclear spin Hamiltonian of AB in the rotating coordinate system is

$$\bar{\mathcal{H}}^{AB} = \sum_s (\omega_{0s} - \omega) I_s^z + \sum_{s>t} J_{s,t} \mathbf{I}_s \cdot \mathbf{I}_t + \omega_1 \sum_s I_s^x, \tag{5-8}$$

where for the moment we do not need to specify which spins belong to A and B. The density matrix equation for AB in the rotating frame between collisions is given as†

$$\dot{\tilde{\rho}}^{AB} = -i\left[\bar{\mathcal{H}}^{AB}, \tilde{\rho}^{AB}\right] + R^{AB}\tilde{\rho}. \tag{5-9}$$

Now rewrite (5-9) as

$$\dot{\tilde{\rho}}^{AB} = -i\left(\bar{\mathcal{L}}_s^{AB} + \bar{\mathcal{L}}_R^{AB}\right)\tilde{\rho}^{AB}, \tag{5-10}$$

†Note we are assuming that the correlation time for molecular motion τ_c is short compared to the exchange time.

where by comparison one sees that

$$\mathcal{L}_s^{AB}\tilde{\rho}^{AB} = \left[\mathcal{H}^{AB}, \tilde{\rho}^{AB}\right], \tag{5-11a}$$

$$i\bar{\mathcal{L}}_R^{AB}\tilde{\rho}^{AB} = R^{AB}\tilde{\rho}. \tag{5-11b}$$

We do this in order to make (5-9) appear as a conventional first-order differential equation. Next, Eq. (5-10) is integrated with the initial condition that

$$\tilde{\rho}^{AB}(t') = \tilde{\rho}^{AB}(\text{col}, t'), \tag{5-11c}$$

where the right-hand side is the value of $\tilde{\rho}^{AB}$ at time t' immediately after a collision which has resulted in the formation of an AB molecule.† The result is that

$$\tilde{\rho}^{AB}(t) = e^{-i(\bar{\mathcal{L}}_s^{AB} + \bar{\mathcal{L}}_R^{AB})(t-t')}\tilde{\rho}^{AB}(\text{col}, t'). \tag{5-12}$$

The probability that after a time $t-t'$ AB will not suffer an exchange collision is

$$\frac{e^{-(t-t')/\tau_{AB}}}{\tau_{AB}} \tag{5-13}$$

and therefore the ensemble average $\tilde{\rho}^{AB}(t)$ is the weighted probability of all AB molecules created at t' which survive until time t

$$\tilde{\rho}^{AB}(t) = \int_{-\infty}^{t} e^{-i(\bar{\mathcal{L}}_s^{AB} + \bar{\mathcal{L}}_R^{AB})(t-t')}\tilde{\rho}^{AB}(\text{col}, t')\frac{e^{-(t-t')/\tau_{AB}}}{\tau_{AB}}\, dt'. \tag{5-14}$$

We next differentiate (5-14) to obtain the differential form of the $\tilde{\rho}^{AB}$ equation. This is done using the result from calculus that for

$$y(t) = \int_A^t g(t, t')\, dt', \tag{5-15}$$

$$\frac{dy}{dt} = \int_A^t \frac{d}{dt}(g(t, t'))\, dt' + g(t, t). \tag{5-16}$$

Applying Eq. (5-16) to (5-14) one obtains the result that

$$\dot{\tilde{\rho}}^{AB} = -i\left(\bar{\mathcal{L}}_s^{AB} + \bar{\mathcal{L}}_R^{AB}\right)\tilde{\rho}^{AB} + \frac{1}{\tau_{AB}}(\tilde{\rho}^{AB}(\text{col}) - \tilde{\rho}^{AB}), \tag{5-17}$$

or going back to the definitions of $\mathcal{L}^{AB}$ and $\bar{\mathcal{L}}_R^{AB}$ in (5-11) we have

$$\dot{\tilde{\rho}}^{AB} = -i\left[\mathcal{H}^{AB}, \tilde{\rho}^{AB}\right] + R^{AB}\tilde{\rho} + (1/\tau_{AB})(\tilde{\rho}^{AB}(\text{col}) - \tilde{\rho}^{AB}). \tag{5-18}$$

To abbreviate the exchange contribution in the density matrix equation we

†See preceding footnote.

write

$$E\tilde{\rho}^{AB} = (1/\tau_{AB})\left[\tilde{\rho}^{AB}(\text{col}) - \tilde{\rho}^{AB}\right], \tag{5-18a}$$

where $\tilde{\rho}$(col) will be evaluated in the following section.

2. Permutation of Indices Method[1]

What remains is to evaluate $\tilde{\rho}^{AB}$(col). We can do this in two ways. First, we can think backward in time, starting with $\tilde{\rho}^{AB}(t)$ *just after* formation of AB by collision and asking what are its anticedents (AC and BD). Alternatively, one can think forward in time taking AC and BD *before* an exchange collision and take them through the chemical reaction. The latter was the first method suggested.[2] However, the former is calculationally, if not conceptually the much simpler approach and the one we will use in this book. This "backward" approach is simpler only if one works in the product representation.

The product representation is written as

$$\psi = \prod_{s=1}^{n} \phi_{ms}, \tag{5-19}$$

where s labels the spins and ϕ_{ms} are eigenfunctions of I^z

$$I_s^z \phi_{ms} = m_s \phi_{ms}. \tag{5-20}$$

For $I = \frac{1}{2}$

$$\phi_{1/2} = \alpha, \qquad \phi_{-(1/2)} = \beta \tag{5-21}$$

and for a molecule made up of only half spins a product wave function would have the form

$$\phi_{\text{molecule}} = \alpha_1 \beta_2 \alpha_3 \beta_4 \beta_5 \alpha_6 \cdots. \tag{5-22}$$

The "product representation" has a special utility because the exchange collision is assumed to happen so fast on the time scale of the nuclear spin Hamiltonian that the nuclear spin wave function is not changed by the collision. This is an example of the so-called "sudden approximation" in quantum mechanics. Thus, the product wave function before and after the collision will be the same.

Now let us see how we make use of these ideas. A particular product wave function of molecule AB can be written as

$$\phi_{ab}^{AB} = \phi_a^A \phi_b^B, \tag{5-22a}$$

where ϕ_a^A and ϕ_b^B are the individual product wave functions of fragments A and B, respectively. An arbitrary matrix element of $\tilde{\rho}^{AB}$ in the product representation *just after* an exchange collision can be written as

$$\langle ab|\tilde{\rho}^{AB}|a'b'\rangle. \tag{5-23}$$

Multiplying expression (5-23) by 1 in the form of

$$1 = \text{Tr}\,\tilde{\rho}^{CD} = \sum_{cd} \langle cd|\tilde{\rho}^{CD}|cd\rangle \tag{5-24}$$

one then has Eq. (5-23) in the form

$$\langle ab|\tilde{\rho}^{AB}|a'b'\rangle = \sum_{cd} \langle ab|\tilde{\rho}^{AB}|a'b'\rangle\langle cd|\tilde{\rho}^{CD}|cd\rangle. \tag{5-25}$$

During an exchange collision there is a mutual reorganization (5-4) where B is replaced by C to form $\tilde{\rho}^{AC}$ and $\tilde{\rho}^{BD}$ and where the product wave functions *ab* and *cd* are rearranged into *ac* and *bd*. Since the product wave function *b* (or *b*′) belongs solely to the fragment B (this would not be true for an eigenfunction), the wave functions, *a*, *b*, *c*, and *d* follow along with the fragments A, B, C, and D, respectively, they are associated with. In other words, the quantum mechanics follows the chemistry; thus Eq. (5-25) becomes after reorganization

$$\langle ab|\tilde{\rho}^{AB}(\text{col})|a'b'\rangle = \sum_{cd} \langle ac|\tilde{\rho}^{AC}|a'c\rangle\langle bd|\tilde{\rho}^{BD}|b'd\rangle, \tag{5-26}$$

which is substituted into (5-18a) to give $(E\tilde{\rho}^{AB})_{ab,a'b}$ as

$$(E\tilde{\rho}^{AB})_{ab,\,a'b'} = \frac{1}{\tau_{AB}}\left[\sum_{cd} \tilde{\rho}^{AC}_{ac,\,a'c}\tilde{\rho}^{BD}_{bd,\,b'd} - \tilde{\rho}^{AB}_{ab,\,a'b'}\right]. \tag{5-26a}$$

To obtain (5-26) from (5-25) exchange B with C in the species superscript and *b* with *c* in the product function label in the subscripts. We call this procedure the **Permutation of Indices** Method (PI). Its great utility is that it follows the *chemical description* of the exchange mechanism. Finally, to repeat, the PI method only applies when the product function is used. The derivation of $\tilde{\rho}$(col) in other representations is very complicated, see below, and lacks the elegant simplicity of the PI method.

The reader is advised not to continue until there is a true understanding of how one proceeds from (5-25) to (5-26).

3. The More Common Exchange Steps and Their $\tilde{\rho}$(col) Functions

Let us now collect a listing of the more commonly encountered exchange steps together with their $\tilde{\rho}$(col) functions.

a. Mutual Exchange of Fragments

$$i \text{ of } \phi_i \qquad \underset{\substack{ab\\a'b'}}{\mathrm{AB}} + \underset{\substack{cd\\c'd'}}{\mathrm{CD}} \underset{k_r}{\overset{k_f}{\rightleftarrows}} \underset{\substack{ac\\a'c'}}{\mathrm{AC}} + \underset{\substack{bd\\b'd'}}{\mathrm{BD}}, \tag{5-27}$$

$$\tilde{\rho}^{\mathrm{CD}}_{cd,\,c'd'}(\mathrm{col}) = \sum_{ab} \tilde{\rho}^{\mathrm{AC}}_{ac,\,a'c}\tilde{\rho}^{\mathrm{BD}}_{bd,\,b'd} \tag{5-27a}$$

$$\tilde{\rho}^{\mathrm{AC}}_{ac,\,a'c'}(\mathrm{col}) = \sum_{bd} \tilde{\rho}^{\mathrm{AB}}_{ab,\,a'b}\tilde{\rho}^{\mathrm{CD}}_{cd,\,c'd}, \tag{5-27b}$$

$$\tilde{\rho}^{\mathrm{BD}}_{bd,\,b'd'}(\mathrm{col}) = \sum_{ac} \tilde{\rho}^{\mathrm{AB}}_{ab,\,ab'}\tilde{\rho}^{\mathrm{CD}}_{cd,\,cd'}, \tag{5-27c}$$

$$\tilde{\rho}^{\mathrm{AB}}_{ab,\,a'b'}(\mathrm{col}) = \sum_{cd} \tilde{\rho}^{\mathrm{AC}}_{ac,\,a'c}\tilde{\rho}^{\mathrm{BD}}_{bd,\,b'd}, \tag{5-27d}$$

$$\underset{\substack{ab\\a'b'}}{\mathrm{AB}} + \underset{\substack{c\\c'}}{\mathrm{B'}} \rightleftarrows \underset{\substack{ac\\a'c'}}{\mathrm{AB'}} + \underset{\substack{b\\b'}}{\mathrm{B}}. \tag{5-28}$$

In Eq. (5-28) B is chemically identical to B′.

$$\tilde{\rho}^{\mathrm{AB}}_{ab,\,a'b'}(\mathrm{col}) = \sum_{c} \tilde{\rho}^{\mathrm{AB}}_{ac,\,a'c}\tilde{\rho}^{\mathrm{B}}_{b,\,b'}, \tag{5-28a}$$

$$\tilde{\rho}^{\mathrm{B}}_{c,\,c'}(\mathrm{col}) = \sum_{ab} \tilde{\rho}^{\mathrm{AB}}_{ac,\,ac'}\tilde{\rho}^{\mathrm{B}}_{b,\,b} = \sum_{a} \tilde{\rho}^{\mathrm{AB}}_{ac,\,ac'}. \tag{5-28b}$$

b. Group Transfer

$$\underset{\substack{ab\\a'b'}}{\mathrm{AB}} + \underset{\substack{cd\\c'd'}}{\mathrm{CD}} \rightleftarrows \underset{\substack{abc\\a'b'c'}}{\mathrm{ABC}} + \underset{\substack{d\\d'}}{\mathrm{D}}, \tag{5-29}$$

$$\tilde{\rho}^{\mathrm{AB}}_{ab,\,a'b'}(\mathrm{col}) = \sum_{cd} \tilde{\rho}^{\mathrm{ABC}}_{abc,\,a'b'c}\tilde{\rho}^{\mathrm{D}}_{d,\,d}$$

$$= \sum_{c} \tilde{\rho}^{\mathrm{ABC}}_{abc,\,a'b'c}, \tag{5-29a}$$

$$\tilde{\rho}^{\mathrm{CD}}_{cd,\,c'd'}(\mathrm{col}) = \sum_{ab} \tilde{\rho}^{\mathrm{ABC}}_{abc,\,abc'}\tilde{\rho}^{\mathrm{D}}_{d,\,d'}, \tag{5-29b}$$

$$\tilde{\rho}^{\mathrm{ABC}}_{abc,\,a'b'c'}(\mathrm{col}) = \sum_{d} \tilde{\rho}^{\mathrm{AB}}_{ab,\,a'b'}\tilde{\rho}^{\mathrm{CD}}_{cd,\,c'd}, \tag{5-29c}$$

$$\tilde{\rho}^{\mathrm{D}}_{d,\,d'}(\mathrm{col}) = \sum_{abc} \tilde{\rho}^{\mathrm{AB}}_{ab,\,ab}\tilde{\rho}^{\mathrm{CD}}_{cd,\,cd'}$$

$$= \sum_{c} \tilde{\rho}^{\mathrm{CD}}_{cd,\,cd'}. \tag{5-29d}$$

c. Dissociation Recombination

$$\underset{\substack{ab\\a'b'}}{AB} \rightleftarrows \underset{\substack{a\\a'}}{A} + \underset{\substack{b\\b'}}{B}, \tag{5-30}$$

$$\tilde{\rho}^{AB}_{ab,\,a'b'}(\text{col}) = \tilde{\rho}^{A}_{a,\,a'}\tilde{\rho}^{B}_{b,\,b'}, \tag{5-30a}$$

$$\tilde{\rho}^{A}_{a,\,a'}(\text{col}) = \sum_b \tilde{\rho}^{AB}_{ab,\,a'b}. \tag{5-30b}$$

d. Unimolecular Rearrangements

These rearrangements include rotation, pseudorotation, Berry rotation, where A is different from B.

$$\underset{\substack{a\\a'}}{A} \rightleftarrows \underset{\substack{a\\a'}}{B}, \tag{5-31}$$

$$\tilde{\rho}^{A}_{a,\,a'}(\text{col}) = \tilde{\rho}^{B}_{a,\,a'}, \tag{5-31a}$$

$$\tilde{\rho}^{B}_{b,\,b'}(\text{col}) = \tilde{\rho}^{A}_{a,\,a'}, \tag{5-31b}$$

e. Unimolecular Degenerate Rearrangements

$$\underset{\substack{abc\\a'b'c'}}{A} \rightleftarrows \underset{\substack{acb\\a'c'b'}}{A}. \tag{5-32}$$

This situation includes all rearrangements where the chemical identity of the species is preserved and the spins become scrambled. In this case the labels of spins are given by the order

$$\tilde{\rho}^{A}_{abc,\,a'b'c'}(\text{col}) = \tilde{\rho}^{A}_{acb,\,a'c'b'}. \tag{5-32a}$$

f. Termolecular Collision

One can envisage inclusion of termolecular steps, say one species is the solvent.

$$\underset{\substack{a\\a'}}{A} + \underset{\substack{b\\b'}}{B} + \underset{\substack{c\\c'}}{C} \rightleftarrows \underset{\substack{abc\\a'b'c'}}{ABC}, \tag{5-33}$$

$$\tilde{\rho}^{A}_{a,\,a'}(\text{col}) = \sum_{bc} \tilde{\rho}^{ABC}_{abc,\,a'bc}, \tag{5-33a}$$

$$\tilde{\rho}_{b,\,b'}(\text{col}) = \sum_{ac} \tilde{\rho}^{ABC}_{abc,\,ab'c}, \tag{5-33b}$$

$$\tilde{\rho}_{c,\,c'}(\text{col}) = \sum_{ab} \tilde{\rho}^{ABC}_{abc,\,abc'}, \tag{5-33c}$$

$$\tilde{\rho}^{ABC}_{abc,\,a'b'c'}(\text{col}) = \tilde{\rho}^{A}_{a,\,a'}\tilde{\rho}^{B}_{b,\,b'}\tilde{\rho}^{C}_{c,\,c'}. \tag{5-33d}$$

4. Systems Undergoing More Than One Exchange Process

The exchange processes considered in the previous section all involved one step and the exchange contributions to the density matrix equations had the form

$$E\tilde{\rho}^{\rm sp} = (1/\tau_{\rm sp})(\tilde{\rho}^{\rm sp}({\rm col}) - \tilde{\rho}^{\rm sp}). \tag{5-34}$$

Now, the question arises as to what is the form of $\tilde{\rho}$(col) in a system undergoing several exchange processes at the same time. The result must clearly be the weighted average

$$\tilde{\rho}^{\rm sp}({\rm col}) = \sum_{\rm ex} \frac{1/\tau_{\rm sp\ ex}}{1/\tau_{\rm sp}} \tilde{\rho}({\rm col\ ex}), \tag{5-35}$$

where ex sums over all exchange processes, $\tau_{\rm sp}$ is the mean lifetime of species sp given as

$$1/\tau_{\rm sp} = \sum_{\rm ex} 1/\tau_{\rm sp\ ex}, \tag{5-36}$$

and $\tau_{\rm sp\ ex}$ is the survival time until sp undergoes a particular exchange step ex. Substituting (5-35) and (5-36) into (5-34) gives that

$$E\tilde{\rho}^{\rm sp} = \sum_{\rm ex} (1/\tau_{\rm sp\ ex})(\tilde{\rho}^{\rm sp}({\rm col\ ex}) - \tilde{\rho}^{\rm sp}). \tag{5-37}$$

As an example, consider the following set of coupled exchange processes

$$\mathrm{AB} \overset{k_a}{\rightleftarrows} \mathrm{A} + \mathrm{B}, \tag{5-38a}$$

$$\mathrm{AB} + \mathrm{C} \overset{k_b}{\rightleftarrows} \mathrm{AC} + \mathrm{D} \tag{5-38b}$$

$$\mathrm{AB} + \mathrm{E} \overset{k_c}{\rightleftarrows} \mathrm{AE} + \mathrm{B}, \tag{5-38c}$$

with

$$1/\tau_{\mathrm{AB},\,a} = k_a, \tag{5-39}$$

$$1/\tau_{\mathrm{AB},\,b} = k_b(\mathrm{C}), \tag{5-40}$$

$$1/\tau_{\mathrm{AB},\,c} = k_c(\mathrm{E}), \tag{5-41}$$

So that $E\tilde{\rho}^{\rm A}B$ is given by

$$\begin{aligned} E\tilde{\rho}^{\rm AB} = {} & k_a(\tilde{\rho}^{\rm AB}({\rm col},\,a) - \tilde{\rho}^{\rm AB}) \\ & + k_b(\mathrm{C})(\tilde{\rho}^{\rm AB}({\rm col},\,b) - \tilde{\rho}^{\rm AB}) \\ & + k_c(\mathrm{E})(\tilde{\rho}^{\rm AB}({\rm col},\,c) - \tilde{\rho}^{\rm AB}) \end{aligned} \tag{5-42}$$

and the different $\tilde{\rho}^{\rm AB}$(col) terms can be readily obtained by use of the catalog in Section 3 of this chapter.

One can see from the above example how the mechanism of the exchange process enters into the density matrix equations in two ways, once in the $1/\tau_{\mathrm{sp\ ex}}$ terms and then in the various $\tilde{\rho}^{\mathrm{sp}}$(col ex) matrix elements. In principle, equations could be written for any exchange process however complicated. However, the cost of computer time at this writing, together with the difficulties of assessing NMR parameters for several species in the same system, places a practical limit on how complicated an exchange process can be usefully handled.

5. The Question of Bilinear Terms in $\tilde{\rho}$(col), SNOB Approximation

For a system undergoing a bimolecular exchange process the $E\tilde{\rho}$ contribution to the density matrix equation could, in principle, contain bilinear terms. To illustrate this question consider the exchange process

$$\mathrm{AB} + \mathrm{B}' \rightleftarrows \mathrm{AB}' + \mathrm{B}. \tag{5-43}$$

As shown in (5-28a) elements of $\tilde{\rho}^{\mathrm{AB}}$(col) are

$$\tilde{\rho}^{\mathrm{AB}}_{ab,\,a'b'}(\mathrm{col}) = \sum_{c} \tilde{\rho}^{\mathrm{AB}}_{ac,\,a'c}\tilde{\rho}^{\mathrm{B}}_{b,\,b'}\,. \tag{5-28a}$$

Equation (5-28a) is manifestly bilinear. However, what we wish to show is that to a high order of precision (corrections will be of order 10^{-5}) we may linearize Eq. (5-28a). This will be automatically taken care of at low power where one writes (see Chapter 4)

$$\tilde{\rho}^{\mathrm{AB}} = \tilde{\rho}^{\mathrm{AB}}_{0} + \tilde{\rho}^{\mathrm{AB}}_{1} \tag{5-44}$$

and keeps terms only linear in $\tilde{\rho}_1$. Thus (5-28a) becomes at low power

$$\tilde{\rho}^{\mathrm{AB}}_{ab,\,a'b'}(\mathrm{col}) = \sum_{c} \left(\tilde{\rho}^{\mathrm{AB}}_{0}\right)_{ac,\,a'c}\left(\tilde{\rho}^{\mathrm{B}}_{1}\right)_{b,\,b'} + \sum_{c} \left(\tilde{\rho}^{\mathrm{AB}}_{1}\right)_{ac,\,a'c}\left(\tilde{\rho}^{\mathrm{B}}_{0}\right)_{b,\,b'} + \text{nonlinear terms}. \tag{5-45}$$

Note that the bilinear zero-order terms

$$\sum_{c} \left(\tilde{\rho}^{\mathrm{AB}}_{0}\right)_{ac,\,a'c}\left(\tilde{\rho}^{\mathrm{B}}_{0}\right)_{b,\,b'} \tag{5-46}$$

are not included in (5-45) as at equilibrium we are guaranteed that

$$\dot{\tilde{\rho}}^{\mathrm{AB}}_{0} = 0 = i\left[\,\bar{\mathcal{L}}^{\mathrm{AB}}_{\mathrm{s}} + \bar{\mathcal{L}}^{\mathrm{AB}}_{\mathrm{R}}\right]\tilde{\rho}^{\mathrm{AB}}_{0} + E\tilde{\rho}^{\mathrm{AB}}_{0}\,. \tag{5-47}$$

Now, consider the high power case where we write

$$\tilde{\rho}^{\mathrm{AB}} = 1/N_{\mathrm{AB}} + \tilde{\rho}^{\mathrm{AB}}_{1'}\,, \tag{5-48}$$

where N_{AB} is the number of nuclear spin states of species AB. Substituting

(5-48) into (5-28a) we obtain

$$\tilde{\rho}_{ab,\,a'b'}^{AB}(\text{col}) = \sum_{\substack{c \\ a=a'}} (1/N_{AB})_{ac,\,a'c}\big(\tilde{\rho}_1^{B}\big)_{b,\,b'}$$

$$+ \sum_{\substack{c \\ b=b'}} (1/N_{B})_{b,\,b'}\big(\tilde{\rho}_1^{B}\big)_{ac,\,a'c}$$

$$+ \sum_{c} \big(\tilde{\rho}_1^{AB}\big)_{ac,\,a'c}\big(\tilde{\rho}_1^{B}\big)_{b,\,b'}\,. \tag{5-49}$$

It has been found that for the usual experimental conditions

$$\big(\tilde{\rho}_1^{AB}\big)_{ac,\,a'c} < 10^{-6}, \qquad a \neq a'.$$

Thus in (5-49) the terms quadratic in ρ will be of order 10^{-12} or less and can be neglected relative to the remaining terms which are of order 10^{-6} or less. The prescription is thus to retain in $\tilde{\rho}(\text{col})$ only those terms linear in $\tilde{\rho}_1$. Note this is not a linearization of the rf field. In practice, it will turn out that when a sum contains terms like

$$\tilde{\rho}_{i,\,i}^{X}\tilde{\rho}_{k,\,l}^{Y} \tag{5-50}$$

and

$$\tilde{\rho}_{i,\,j}^{X}\tilde{\rho}_{k,\,l}^{Y} \tag{5-51}$$

the latter can be neglected and the former written as

$$\tilde{\rho}_{i,\,i}^{X}\tilde{\rho}_{k,\,l}^{Y} \sim (1/N_x)\tilde{\rho}_{k,\,l}^{Y}\,. \tag{5-52}$$

We call this Selective Neglect of Bilinear terms (SNOB).

6. Alternative Evaluation of $\tilde{\rho}(\text{col})$, Forward in Time

As mentioned above, we can also evaluate $\tilde{\rho}^{AB}(\text{col})$ by looking at the process from a forward time viewpoint. Thus a collision

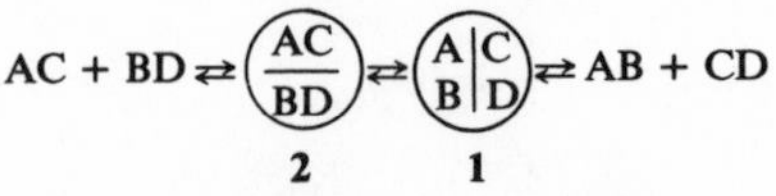

$$AC + BD \rightleftarrows \left(\frac{AC}{BD}\right) \rightleftarrows \left(\begin{array}{c|c} A & C \\ B & D \end{array}\right) \rightleftarrows AB + CD \tag{5-53}$$
$$ \mathbf{2} \phantom{\left(\frac{AC}{BD}\right) \rightleftarrows} \mathbf{1}$$

forms a complex **2** of AC with BD with the density matrix

$$\tilde{\rho}^{AC} \times \tilde{\rho}^{BD}. \tag{5-54}$$

After rearrangement to complex **1**, but before dissociation to AB and CD, the density matrix is given by

$$U\tilde{\rho}^{AC} \times \tilde{\rho}^{BD}U^{-1}, \tag{5-55}$$

where U is the reorganization operator defined below. Finally, the rearranged complex **1** dissociates and we separate out the AB part of (5-55) by taking the trace over the states of CD written as

$$\tilde{\rho}^{AB}(\mathrm{col}) = \mathop{\mathrm{Tr}}_{cd} \mathbf{U}\tilde{\rho}^{AC} \times \tilde{\rho}^{BD}\mathbf{U}^{-1}. \qquad (5\text{-}56)$$

The reorganization operator **U** is defined to transform wave functions of the AC · BD complex **2** into wave functions of the AB · CD complex **1**. Thus, in the eigenfunction representation **U** is defined by

$$\mathbf{U}\psi_n^{AC}\psi_l^{BD} = \sum_{i,j} c_{i,j}\psi_i^{AB}\psi_j^{CD}, \qquad (5\text{-}57)$$

where ψ_i^{AB}, etc., are nuclear spin eigenfunctions for the different species present. In the eigenfunction representation the elements of **U** turn out to be complicated functions of the NMR parameters and have to be obtained, in general, by numerical computer techniques. On the other hand, in the product representation we have

$$\mathbf{U}\phi_n^{AC}\phi_l^{BD} = \phi_i^{AB}\phi_j^{CD}; \qquad (5\text{-}58)$$

note there is no sum in (5-58). This is because every product wave function of the AC · BD complex is one and only one product wave function of the AB · CD complex. Thus, in the product representation, **U** is a permutation matrix with elements zero or one.

Let us work out an example using the forward procedure. Consider a mutual exchange of protons H_a and H_b between two different environments A and B.

State labels:

$$\underset{a}{AH_a} + \underset{b}{BH_b} \rightleftarrows \underset{b}{AH_b} + \underset{a}{BH_a}. \qquad (5\text{-}59)$$

The cross product density matrix for the exchanging system in the product representation (states α and β) is given as

$$\tilde{\rho}^A \times \tilde{\rho}^B = \begin{bmatrix} \tilde{\rho}^A_{\alpha,\alpha}\cdot\tilde{\rho}^B_{\alpha,\alpha} & \tilde{\rho}^A_{\alpha,\alpha}\cdot\tilde{\rho}^B_{\alpha,\beta} & \tilde{\rho}^A_{\alpha,\beta}\cdot\tilde{\rho}^B_{\alpha,\alpha} & \tilde{\rho}^A_{\alpha,\beta}\cdot\tilde{\rho}^B_{\alpha,\beta} \\ \tilde{\rho}^A_{\alpha,\alpha}\cdot\tilde{\rho}^B_{\beta,\alpha} & \tilde{\rho}^A_{\alpha,\alpha}\cdot\tilde{\rho}^B_{\beta,\beta} & \tilde{\rho}^A_{\alpha,\beta}\cdot\tilde{\rho}^B_{\beta,\alpha} & \tilde{\rho}^A_{\alpha,\beta}\cdot\tilde{\rho}^B_{\beta,\beta} \\ \tilde{\rho}^A_{\beta,\alpha}\cdot\tilde{\rho}^B_{\alpha,\alpha} & \tilde{\rho}^A_{\beta,\alpha}\cdot\tilde{\rho}^B_{\alpha,\beta} & \tilde{\rho}^A_{\beta,\beta}\cdot\tilde{\rho}^B_{\alpha,\alpha} & \tilde{\rho}^A_{\beta,\beta}\cdot\tilde{\rho}^B_{\alpha,\beta} \\ \tilde{\rho}^A_{\beta,\alpha}\cdot\tilde{\rho}^B_{\beta,\alpha} & \tilde{\rho}^A_{\beta,\alpha}\cdot\tilde{\rho}^B_{\beta,\beta} & \tilde{\rho}^A_{\beta,\beta}\cdot\tilde{\rho}^B_{\beta,\alpha} & \tilde{\rho}^A_{\beta,\beta}\cdot\tilde{\rho}^B_{\beta,\beta} \end{bmatrix} \qquad (5\text{-}60)$$

the rows and columns being labeled by the product states $\phi_a^A\phi_b^B$ and $\phi_{a'}^A\phi_{b'}^B$, respectively. Note that in Eq. (5-60) we can recover $\tilde{\rho}^A$ by taking the trace over b, i.e.,

$$\tilde{\rho}^A_{a,a'} = \mathop{\mathrm{Tr}}_{b}\langle a|\tilde{\rho}^A \times \tilde{\rho}^B|a'\rangle$$

$$= \sum_b \langle ab|\tilde{\rho}^A \times \tilde{\rho}^B|a'b\rangle. \qquad (5\text{-}61)$$

This is illustrated by the outlined terms in (5-60), thus

$$\begin{aligned}\tilde{\rho}^{A}_{\alpha,\beta} &= \tilde{\rho}^{A}_{\alpha,\beta}\tilde{\rho}^{B}_{\alpha,\alpha} + \tilde{\rho}^{A}_{\alpha,\beta}\tilde{\rho}^{B}_{\beta,\beta} \\ &= \tilde{\rho}^{A}_{\alpha,\beta}\left(\tilde{\rho}^{B}_{\alpha,\alpha} + \tilde{\rho}^{B}_{\beta,\beta}\right) = \tilde{\rho}^{A}_{\alpha,\beta} \end{aligned} \tag{5-62}$$

as

$$\operatorname{Tr} \tilde{\rho}^{B} = 1 = \tilde{\rho}^{B}_{\alpha,\alpha} + \tilde{\rho}^{B}_{\beta,\beta}\,. \tag{5-63}$$

The reorganization operator has the effect of exchanging a hydrogen on A with one on B, thus, for example,

$$\mathbf{U}\alpha^{A}\beta^{B} = \beta^{A}\alpha^{B} \tag{5-64}$$

and the reorganization matrix $\mathbf{U}$ acting on a column of product states appears as

$$\begin{bmatrix} 1 & 0 & 0 & 0 \\ 0 & 0 & 1 & 0 \\ 0 & 1 & 0 & 0 \\ 0 & 0 & 0 & 1 \end{bmatrix}\begin{bmatrix} \alpha\alpha \\ \alpha\beta \\ \beta\alpha \\ \beta\beta \end{bmatrix} = \begin{bmatrix} \alpha\alpha \\ \beta\alpha \\ \alpha\beta \\ \beta\beta \end{bmatrix}. \tag{5-65}$$

Then, carrying out the multiplications in $\mathbf{U}\tilde{\rho}^{A} \times \tilde{\rho}^{B}\mathbf{U}^{-1}$ gives

$$\mathbf{U}\tilde{\rho}^{A} \times \tilde{\rho}^{B}\mathbf{U}^{-1} = \begin{bmatrix} \tilde{\rho}^{A}_{\alpha,\alpha}\cdot\tilde{\rho}^{B}_{\alpha,\alpha} & \tilde{\rho}^{A}_{\alpha,\beta}\cdot\tilde{\rho}^{B}_{\alpha,\alpha} & \boxed{\tilde{\rho}^{A}_{\alpha,\alpha}\cdot\tilde{\rho}^{B}_{\alpha,\beta}} & \tilde{\rho}^{A}_{\alpha,\beta}\cdot\tilde{\rho}^{B}_{\alpha,\beta} \\ \tilde{\rho}^{A}_{\beta,\alpha}\cdot\tilde{\rho}^{B}_{\alpha,\alpha} & \tilde{\rho}^{A}_{\beta,\beta}\cdot\tilde{\rho}^{B}_{\alpha,\alpha} & \tilde{\rho}^{A}_{\beta,\alpha}\cdot\tilde{\rho}^{B}_{\alpha,\beta} & \boxed{\tilde{\rho}^{A}_{\beta,\beta}\cdot\tilde{\rho}^{B}_{\alpha,\beta}} \\ \tilde{\rho}^{A}_{\alpha,\alpha}\cdot\tilde{\rho}^{B}_{\beta,\alpha} & \tilde{\rho}^{A}_{\alpha,\beta}\cdot\tilde{\rho}^{B}_{\beta,\alpha} & \tilde{\rho}^{A}_{\alpha,\alpha}\cdot\tilde{\rho}^{B}_{\beta,\beta} & \tilde{\rho}^{A}_{\alpha,\beta}\cdot\tilde{\rho}^{B}_{\beta,\beta} \\ \tilde{\rho}^{A}_{\beta,\alpha}\cdot\tilde{\rho}^{B}_{\beta,\alpha} & \tilde{\rho}^{A}_{\beta,\beta}\cdot\tilde{\rho}^{B}_{\beta,\alpha} & \tilde{\rho}^{A}_{\beta,\alpha}\cdot\tilde{\rho}^{B}_{\beta,\beta} & \tilde{\rho}^{A}_{\beta,\beta}\cdot\tilde{\rho}^{B}_{\beta,\beta} \end{bmatrix}, \tag{5-66}$$

where the rows and columns carry the same labels as the $\tilde{\rho}^{A} \times \tilde{\rho}^{B}$ matrix [(5-60)]. Thus, the

$$\left[\mathbf{U}\tilde{\rho}^{A} \times \tilde{\rho}^{B}\mathbf{U}^{-1}\right]_{\alpha\beta,\beta\beta} \tag{5-67}$$

element is given by

$$\tilde{\rho}^{A}_{\beta,\beta}\tilde{\rho}^{B}_{\alpha,\beta}\,. \tag{5-68}$$

To obtain the element $(\tilde{\rho}^{A}(\text{col}))_{\alpha,\beta}$ we take as before the terms outlined in (5-66) as

$$\begin{aligned}(\tilde{\rho}^{A}(\text{col}))_{\alpha,\beta} &= \tilde{\rho}^{A}_{\alpha,\alpha}\tilde{\rho}^{B}_{\alpha,\beta} + \tilde{\rho}^{A}_{\beta,\beta}\tilde{\rho}^{B}_{\alpha,\beta} \\ &= \tilde{\rho}^{B}_{\alpha,\beta}. \end{aligned} \tag{5-69}$$

In general, one then finds

$$\tilde{\rho}^{A}(\text{col}) = \tilde{\rho}^{B}. \tag{5-70}$$

The result in (5-70) can also be obtained using the rules of matrix algebra, without explicitly writing down the reorganization matrix at all. A nonzero element of $\mathbf{U}\tilde{\rho}^{A} \times \tilde{\rho}^{B}\mathbf{U}^{-1}$ is

$$\begin{aligned}\left[\mathbf{U}\tilde{\rho}^{A} \times \tilde{\rho}^{B}\mathbf{U}^{-1}\right]_{ab,\,a'b'} &= \langle ab|\mathbf{U}|ba\rangle\langle ba|\tilde{\rho}^{A} \times \tilde{\rho}^{B}|b'a'\rangle\langle b'a'|\mathbf{U}|a'b'\rangle \\ &= \langle ba|\tilde{\rho}^{A} \times \tilde{\rho}^{B}|b'a'\rangle \\ &= \langle b|\tilde{\rho}^{A}|b'\rangle\langle a|\tilde{\rho}^{B}|a'\rangle \end{aligned} \tag{5-71}$$

as

$$|\mathbf{U}|ba = ab. \tag{5-72}$$

Then, an element of $\tilde{\rho}^{A}(\text{col})$ is

$$\langle a|\tilde{\rho}^{A}(\text{col})|a'\rangle = \langle a|\text{Tr}_{b}\, \mathbf{U}\tilde{\rho}^{A} \times \tilde{\rho}^{B}\mathbf{U}^{-1}|a'\rangle, \tag{5-73}$$

where Tr_{b} means sum over all elements where $b = b'$. This gives the final result, using (5-71) for an element of $\tilde{\rho}^{A}(\text{col})$ that

$$\langle a|\tilde{\rho}^{A}(\text{col})|a'\rangle = \sum_{b} \langle b|\tilde{\rho}^{A}|b\rangle\langle a|\tilde{\rho}^{B}|a'\rangle = \langle a|\tilde{\rho}^{B}|a'\rangle, \tag{5-74}$$

which is identical to that obtained above (for $a = \alpha$, $a' = \beta$) and by the PI method described previously in Section 2 of this chapter. In fact, the $\tilde{\rho}(\text{col})$ terms for all exchange processes (see Section 3) can be derived either in the general manner just described or by the PI method.

7. Summary

The effect of exchange processes on the density matrix equations is shown to lead to a term of the form $-(1/\tau)(\rho - \rho(\text{col}))$. Two general methods are described for evaluating $\tilde{\rho}(\text{col})$. In particular, it is shown how in the product representation one of these methods, called Permutation of Indices (PI), conveniently evaluates the $\tilde{\rho}(\text{col})$ elements. It also follows the chemical description of the exchange mechanism. A table is given of $\tilde{\rho}(\text{col})$ terms for most experimentally encountered exchange steps.

Problems

1. Show how the Bloch equations for two spins, X and X^*, exchanging

$$AX^* + BX \rightleftarrows AX + BX^*$$

between two environments, A and B, can be derived from the density matrix.

2. (a) Evaluate $E\tilde{\rho}^{AB}$ and $E\tilde{\rho}^{B}$ (5-34) for the system

$$AB + B^* \rightleftarrows AB^* + B,$$

where B and B* are identical, A is strongly coupled to B(B*), and I for A and B(B*) is $\frac{1}{2}$.

(b) Obtain a general expression for the elements $\tilde{\rho}(\mathrm{col})$ for species AB. Compare results with listings given in Section 3.

3. Show that a Bloch equation cannot be derived for the system defined by

$$\mathcal{H} = \omega_{0A} I_A^Z + \omega_{0B} + J I_A \cdot I_B.$$

REFERENCES

1. J. I. Kaplan and G. Fraenkel, *J. Amer. Chem. Soc.* **94**, 2907 (1972).
2. J. I. Kaplan, *J. Chem. Phys.* **28**, 278 (1958); **40**, 462 (1964)
3. S. Alexander, *J. Chem. Phys.* **37**, 967 (1962); **38**, 1787 (1963); **40**, 2741 (1964).
4. E. Grunwald, A. Loewenstein, and S. Meiboom, *J. Chem. Phys.* **27**, 630 (1957).

Chapter VI

NMR LINESHAPES FOR EXCHANGING SYSTEMS UNDER CONDITIONS OF LOW rf POWER

1. Introduction

In this chapter we will discuss the low power NMR absorption for exchanging systems, i.e., the absorption calculated keeping only those terms linear in the rf field.[1] Experimentally the limiting low power, or linearized, absorption can be obtained either by cw experiments or by fourier transform (ft) of the free induction decay. The latter technique (discussed in Chapter 9) is only useful if the spread of the spectrum is such that the 90° pulse can act equally on all transitions. In what follows we will discuss everything in terms of a cw calculation, but of course everything we say would apply equally to the ft technique.

The first order of business will be to indicate the form of the density matrix equation after linearization. There will follow an analytical investigation of exchange between two uncoupled sites and its extension to exchange among many half-spin sites (Kubo–Anderson–Sack). Next, it will be shown how to treat equivalent spins exactly, for instance, the three protons of a methyl group. This treatment is applied to exchange among multi-half-spin sites occupied by magnetically equivalent spins, e.g., methyl groups. It will be shown how the density matrix treatment leads, after equivalence factoring, to a set of pseudo half-spin lineshape equations. There follows an account of the varying degrees of approximation with which lineshape problems can be handled.

In Section 4 effects due to relaxation are discussed followed by, in Section 5, the particular case of quadrupole effects and how they influence NMR lineshapes. There are some remarks on effects due to low concentration species. Lineshapes from solutes in the nematic phase are described. To conclude there is mention of experimental questions such as temperature control, field homogeneity, and the extraction of intrinsic NMR parameters.

Finally—a word on notation. For simplicity and to reduce clutter we shall write $\tilde{\rho}$ to mean $\tilde{\rho}_s$, the nuclear spin density matrix.

2. Final Form of the Density Matrix Equation at Low rf Power

a. *Evaluation of* $[\overline{\mathcal{H}}_{rf}, \tilde{\rho}]$

The density matrix equation, as obtained in the preceding chapters, is given as

$$\dot{\rho} = -i[\mathcal{H}_0, \rho] - i[\mathcal{H}_{\mathrm{rf}}, \rho] + R\rho + E\rho, \tag{6-1}$$

where R and E are the relaxation and exchange operators, respectively, assuming a circularly polarized rf field. Going into the rotating coordinate system at frequency ω of the rf field (6-1) appears as

$$\dot{\tilde{\rho}} = 0 = -i[\overline{\mathcal{H}}_0, \tilde{\rho}] - i\sum_s \omega\!\!\!/_{1s}[I_s^x, \tilde{\rho}] + R\tilde{\rho} + E\tilde{\rho}, \tag{6-2}$$

where s labels nuclei and the other symbols have their usual meanings. Linearization consists of writing

$$\tilde{\rho} = \rho_0 + \tilde{\rho}_1, \tag{6-3}$$

where ρ_0 applies to the equilibrium distribution. Substituting (6-3) into (6-2) and keeping only those terms linear in the rf field yields

$$0 = -i[\overline{\mathcal{H}}_0, \tilde{\rho}_1] - i\sum_s \omega\!\!\!/_{1s}[I_s^x, \rho_0] + R\tilde{\rho}_1 + E\tilde{\rho}_1, \tag{6-4}$$

where we have made use of the result that for equilibrium

$$-i[\mathcal{H}_0, \rho_0] + R\rho_0 + E\rho_0 = 0. \tag{6-5}$$

For conditions typical of those used in NMR experiments we have

$$(\bar{\omega\!\!\!/}_1 - \omega\!\!\!/_{1s})/\bar{\omega\!\!\!/}_1 \ll 1, \tag{6-6}$$

$$(\bar{\omega}_0 - \omega_{0s})/\bar{\omega}_0 \ll 1, \tag{6-7}$$

$$J_{s,t}/\bar{\omega}_0 \ll 1, \tag{6-8}$$

$$\hbar\bar{\omega}_0/kT \ll 1, \tag{6-9}$$

where s, t stand for spins, ω_{0s} and ω_{1s} are the resonance frequency, and rf field strength, respectively, for a particular nucleus s, and the bars signify average values. Relationships (6-6) and (6-7) state that resonance frequency differences are small compared to the average as are rf coupling strengths. Using (6-8) and (6-9) one expands ρ_0 in (6-4) as

$$\rho_0 \simeq \frac{\mathcal{I} - (\hbar/kT)\Sigma_s \omega_{0s} I_s^z}{N}, \tag{6-10}$$

which allows us to write

$$- i\sum_s \omega_{1s}[I_s^x, \rho_0] \tag{6-11}$$

as

$$(i\hbar/NkT)\sum_s \omega_{1s}\omega_{0s}[I_s^x, I_s^z] = (\hbar/NkT)\sum \omega_{1s}\omega_{0s} I_s^y, \tag{6-12}$$

where we have used the commutator relationship

$$[I^x, I^z] = -iI^y. \tag{6-13}$$

A further simplification can be now introduced. Relative intensities of NMR absorption peaks can be measured to no better than $\frac{1}{10^3}$. Since the absorption is proportional to elements

$$- i\left[\mathcal{H}_{rf}, \rho_0\right]_{i,k},$$

then the summation over $\omega_{1s}\omega_{0s}$ in (6-12) can be replaced by an average (this will not be true for different isotopes) so that

$$-i\left[\overline{\mathcal{H}}_{rf}\rho_0\right] \simeq (\hbar\bar{\omega}_1\bar{\omega}_0/NkT)\sum_s I_s^y, \tag{6-14}$$

which may be abbreviated

$$- i[\mathcal{H}_{rf}, \rho_0] = (2\beta'/N)\sum_s I_s^y. \tag{6-15}$$

b. The NMR Absorption

Last, we need an expression for the absorption. This is given in the rotating coordinate frame by the rf field $\bar{\omega}_1$ times the component of magnetization out of phase with the rf field and weighted with concentration factors (sp), as [note the rf field in the rotating coordinate system is in the x direction (6-4)]

$$\text{Abs} \sim \sum_{\text{sp}} \sum_s (\text{sp})\bar{\omega}_{1,\text{sp}} \text{Tr}\, I_{s,\text{sp}}^y \tilde{\rho}^{\text{sp}}. \tag{6-16}$$

One evaluates the trace part of (6-16) by writing

$$I^y = -\tfrac{1}{2}i(I^+ - I^-) \tag{6-17}$$

and noting that

$$I^+|m\rangle = \sqrt{I(I+1) - m(m+1)}\,|m+1\rangle, \tag{6-18}$$

$$I^-|m\rangle = \sqrt{I(I+1) - m(m-1)}\,|m-1\rangle. \tag{6-19}$$

Using Eqs. (6-17) and (6-18) one evaluates

$$\mathrm{Tr}\ \tilde{\rho} I^y \tag{6-20}$$

in the product representation as

$$\begin{aligned} \mathrm{Tr}\ I^y\tilde{\rho} &= -\frac{i}{2}\sum_m \langle m|\tilde{\rho}(I^+ - I^-)|m\rangle \\ &= -\frac{i}{2}\sum_m \{\langle (m-1)|\tilde{\rho}|m\rangle \sqrt{I(I+1) - m(m-1)} \\ &\quad - \langle m|\tilde{\rho}|(m-1)\rangle \sqrt{I(I+1) - m(m-1)}\,\}. \end{aligned} \tag{6-21}$$

Now, from the hermitian property of $\tilde{\rho}$ we have that

$$\langle m|\tilde{\rho}|m'\rangle = \langle m'|\tilde{\rho}|m\rangle^*. \tag{6-22}$$

Substituting (6-22) into (6-21) one then obtains

$$\begin{aligned} \mathrm{Tr}\ I^{y'}\tilde{\rho} &= -\frac{i}{2}\sum_m \sqrt{I(I+1) - m(m-1)} \\ &\quad \times \left[-\langle m|\tilde{\rho}|(m-1)\rangle + \langle m|\tilde{\rho}|(m-1)^*\rangle\right] \\ &= -\sum_m \mathrm{Im}\langle m|\tilde{\rho}|(m-1)\rangle \sqrt{I(I+1) - m(m-1)}\,, \end{aligned} \tag{6-23}$$

where Im means "imaginary part of".

3. Exchanging Systems

a. Exchange between Two Sites, Uncoupled[2–4]

(1) Introduction

It is constructive to start off by considering the two site uncoupled exchange case since it can be solved analytically. In this kind of exchanging system there are two sites, each occupied by a single proton or by groups of magnetically equivalent protons (e.g., methyl or *t*-butyl groups). The nuclei at each site give rise to single lines. The two sites could be within the same molecule or on different molecules as shown in (6-24) to (6-27)

$$AH \rightleftarrows BH, \tag{6-24}$$

$$A + BH \rightleftarrows B + AH, \tag{6-25}$$

$$AHBH^* \rightleftarrows AH^*BH, \tag{6-26}$$

$$AH + BH^* \rightleftarrows AH^* + BH, \tag{6-27}$$

where exchange of hydrogens or equivalent groups of hydrogens by rearrangement or bimolecular reaction averages the two resonances. One of the first examples of a rate process investigated with NMR was hindered rotation in N,N-dimethylformamide.

(6-28)

In all these cases the group of hydrogens occupying one site is simulated as a single spin, $\frac{1}{2}$ (we will shortly treat equivalency).

We will now treat intramolecular rearrangement in the two site case

$$A \underset{k_{-1}}{\overset{k_1}{\rightleftarrows}} B, \qquad (B)/(A) = k_1/k_{-1}, \tag{6-29}$$

where A and B each contain one proton or a group of equivalent protons not coupled to other nuclei. The parameters relevant to this calculation (for two sites, spin $\frac{1}{2}$, not coupled) are collected in (6-30) where T_t and T_l stand for transverse and longitudinal relaxation times, ω_{0A} and ω_{0B} are shifts, and τ's mean preexchange lifetimes.

Rate constants	Random field relaxation	Shifts
$1/\tau_A = k_1$,	T_{tA}, T_{1A}	ω_{0A}
$1/\tau_B = k_{-1}$,	T_{tB}, T_{1B}	ω_{0B}

$$\text{Spin functions:} \quad \phi_i^A = \begin{cases} \alpha, & i = 1, \\ \beta, & i = 2, \end{cases} \qquad \phi_i^B = \begin{cases} \alpha, & i = 1 \\ \beta, & i = 2 \end{cases}. \tag{6-30}$$

As described in Chapter V, the exchange contribution to the density matrix equation is

$$\left[E\tilde{\rho}_1^A\right]_{1,2} = (1/\tau_A)\left[(\tilde{\rho}_1^B)_{1,2} - (\tilde{\rho}_1^A)_{1,2}\right], \tag{6-31}$$

$$\left[E\tilde{\rho}_1^B\right]_{1,2} = (1/\tau_B)\left[(\tilde{\rho}_1^A)_{1,2} - (\tilde{\rho}_1^B)_{1,2}\right]. \tag{6-32}$$

Since we need $(\tilde{\rho}_1^A)_{1,2}$ and $(\tilde{\rho}_1^B)_{1,2}$ to solve for the absorption, the required elements of Eq. (6-4) are

$$0=\langle 1|-i\left[\mathcal{H}^{A}{}_{0},\tilde{\rho}_1^{A}\right]+(\beta'/2)I_A^y+R\tilde{\rho}+(\tilde{\rho}_1^{B}-\tilde{\rho}_1^{A})/\tau_A|2\rangle, \qquad (6\text{-}33)$$

where $R\tilde{\rho}$ is given in (5-2) and

$$\beta' = \hbar\bar{\omega}_0\,\bar{\omega}_1/2kT. \qquad (6\text{-}34)$$

A second equation is obtained from it by replacing B with A and A with B wherever they occur. Then using as an example the single spin problem as illustrated in Chapter III, the two equations we need are found to be

$$\left[i(\omega-\omega_{0A})-(2/T_{tA})-(1/T_{1A})-(1/\tau_A)\right](\tilde{\rho}_1)_{1,2} + (\tilde{\rho}_1^{B})_{1,2}(1/\tau_A)=i\beta'/2, \qquad (6\text{-}35)$$

$$\left[i(\omega-\omega_{0B})-(2/T_{tB})-(1/\tau_B)\right](\tilde{\rho}_1^{B})_{1,2} + \tilde{\rho}_{1,2}^{A}(1/\tau_B)=+i\beta'/2. \qquad (6\text{-}36)$$

It is convenient to simplify the relaxation terms using

$$1/T_A = (1/T_{t\,A}) + (1/T_{1\,A}) \qquad (6\text{-}37)$$

and a similar equation for $1/T_B$,

$$1/T_B = (1/T_{tB}) + (1/T_{1B}). \qquad (6\text{-}38)$$

The coupled equations in matrix form appear as

$$\begin{bmatrix} -i(\omega_{0A}-\omega)-(1/T_A)-(1/\tau_A) & 1/\tau_A \\ 1/\tau_B & -(\omega_{0B}-\omega)-(1/T_B)-(1/\tau_B) \end{bmatrix} \begin{bmatrix} (\tilde{\rho}_1^{A})_{1,2} \\ (\tilde{\rho}_1^{B})_{1,2} \end{bmatrix} = \frac{i\beta'}{2}\begin{bmatrix} 1 \\ 1 \end{bmatrix}. \qquad (6\text{-}39)$$

The NMR absorption, given as

$$\text{Abs} \sim (\text{A})\text{Tr}\,\tilde{\rho}^{A}I_A^y + (\text{B})\text{Tr}\,\tilde{\rho}^{B}I_B^y, \qquad (6\text{-}40)$$

is calculated to be

$$\text{Abs} \sim -(\text{A})\text{Im}(\tilde{\rho}_1^{A})_{1,2} - (\text{B})\text{Im}(\tilde{\rho}_1^{B})_{1,2}, \qquad (6\text{-}41)$$

using (6-23) above.

The solution of (6-39) obtained analytically in the standard way is

$$\begin{bmatrix} (\tilde{\rho}_1^{A})_{1,2} \\ (\tilde{\rho}_1^{B})_{1,2} \end{bmatrix} = \frac{i\beta'}{2D}\begin{bmatrix} -i(\omega_{0B}-\omega)-(1/T_B)-(1/\tau_B) & -1/\tau_A \\ -1/\tau_B & -i(\omega_{0A}-\omega)-(1/T_A)-(1/\tau_A) \end{bmatrix}\begin{bmatrix} 1 \\ 1 \end{bmatrix}, \qquad (6\text{-}42)$$

where D is the determinant of the coefficient matrix in (6-39).

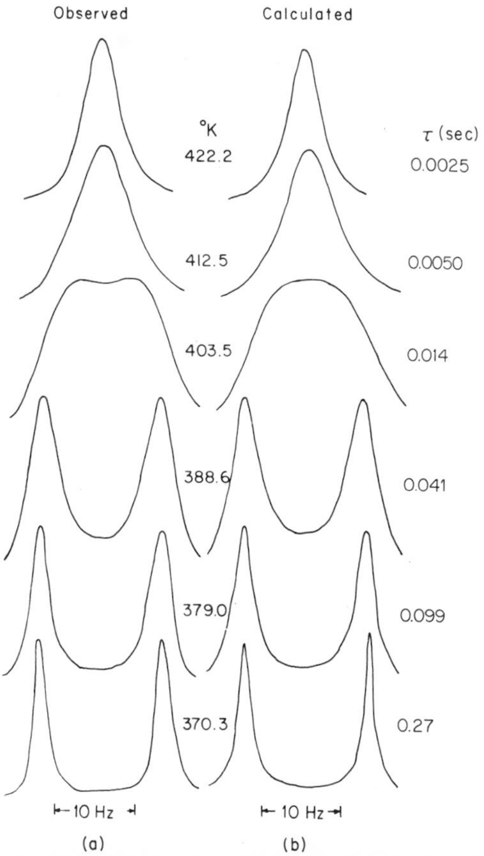

Fig. 6-1. (a) 100-MHz proton NMR of *N,N*-dimethylformamide-D_1 in carbon tetrachloride as a function of temperature, (b) calculated lineshapes. [From W. C. Tung, Ph.D. Thesis, Michigan State Univ., East Lansing, 1968; *Diss. Abstr.* **30**, 162B (1968).]

An example where experimental and theoretical *N*-methyl NMR lineshapes for *N,N*-dimethylformamide-D_1 are compared is given in Fig. 6-1.

Much can be determined analytically about the two site case, especially under conditions of limiting slow and fast exchange. For instance, if we are solely interested in resonance positions and linewidths and not amplitudes we only need to obtain the zeros of D.

(2) Slow Exchange

For slow exchange the off diagonal elements $1/\tau_A$ and $1/\tau_B$ can be ignored. Then the solution, given in the form

$$\tilde{\rho} = (\mathcal{I} i\omega + \mathrm{A})^{-1}\mathrm{B}, \tag{6-43}$$

becomes

$$\begin{bmatrix} (\tilde{\rho}_1^{\mathrm{A}})_{1,2} \\ (\tilde{\rho}_1^{\mathrm{B}})_{1,2} \end{bmatrix} = \frac{i\beta'}{2} \begin{bmatrix} \dfrac{1}{-i(\omega_{0\mathrm{A}}-\omega)-(1/T_{\mathrm{A}})-(1/\tau_{\mathrm{A}})} & 0 \\ 0 & \dfrac{1}{-i(\omega_{0\mathrm{B}}-\omega)-(1/T_{\mathrm{B}})-(1/\tau_{\mathrm{B}})} \end{bmatrix} \begin{bmatrix} 1 \\ 1 \end{bmatrix} \tag{6-44}$$

By inspection of (6-44) the zeros are seen to be

$$-i(\omega_{0\mathrm{A}}-\omega) = (1/T_{\mathrm{A}}) + (1/\tau_{\mathrm{A}}), \qquad -i(\omega_{0\mathrm{B}}-\omega) = (1/T_{\mathrm{B}}) + (1/\tau_{\mathrm{B}}), \tag{6-45}$$

and we thus obtain two resonances at $\omega_{0\mathrm{A}}$ and $\omega_{0\mathrm{B}}$ with widths given as

$$d\omega_{\mathrm{A}} = (1/T_{\mathrm{A}}) + (1/\tau_{\mathrm{A}}), \qquad d\omega_{\mathrm{B}} = (1/T_{\mathrm{B}}) + (1/\tau_{\mathrm{B}}). \tag{6-46}$$

(3) Intermediate Exchange

For conditions of intermediate exchange one has to use the solution of (6-39) in its complete form. Various analytic expressions for the absorption have been published elsewhere.[2-4] They are cumbersome to use and not really necessary since the problem can easily be handled with a minicomputer.

(4) Fast Exchange

When exchange is very fast compared to the NMR time scale, the A and B resonances become completely averaged and a single line is obtained of width

$$d\omega_{\mathrm{fast}} = (p_{\mathrm{A}}/T_{\mathrm{A}}) + (p_{\mathrm{B}}/T_{\mathrm{B}}), \tag{6-47}$$

which is the average width of the two original lines (p_{A} and p_{B} are fractional populations). Just short of the fast exchange limit the line is somewhat broader, the width being given by

$$d\omega = (p_{\mathrm{A}}/T_{\mathrm{A}}) + (p_{\mathrm{B}}/T_{\mathrm{B}}) + p_{\mathrm{A}}^2 p_{\mathrm{B}}^2 \delta^2 \tau, \tag{6-48}$$

where δ is the chemical shift $\omega_{0\mathrm{A}} - \omega_{0\mathrm{B}}$. It is instructive to derive this result since the procedure is not well known.

When the denominator D in Eq. (6-42) goes to zero there is an absorption maximum. D, the determinant of the matrix in Eq. (6-39), is given as

$$D = 0 = [-i(\omega_{0A} - \omega) - (1/T_A) - (1/\tau_A)] \times [-i(\omega_{0B} - \omega) - (1/T_B) - (1/\tau_B)] - (1/\tau_A\tau_B). \quad (6\text{-}49)$$

First, make the following substitutions

$$1/\tau = p_A/\tau_A = p_B/\tau_B, \quad (6\text{-}50)$$

$$\omega_0 = p_A\omega_{0A} + p_B\omega_{0B}, \quad (6\text{-}51)$$

$$\delta = \omega_{0A} - \omega_{0B}, \quad (6\text{-}52)$$

$$\Delta = \omega_0 - \omega, \quad (6\text{-}53)$$

$$\omega_{0A} - \omega = \Delta + p_B\delta, \qquad \omega_{0B} - \omega = \Delta - p_A\delta. \quad (6\text{-}54)$$

Substituting for ω_{0A}, ω_{0B}, ω, $1/\tau_A$, and $1/\tau_B$ in (6-49) gives

$$0 = \Delta^2 - \Delta\delta(p_A - p_B) + i[(1/T_A) + (1/T_B) + (1/\tau p_A p_B)]\Delta - p_A p_B \delta^2 - i\delta[(p_B/T_B) - (p_A/T_A)] - (1/T_A T_B) - (1/\tau)[(1/p_A T_A) + (1/p_B T_B)]. \quad (6\text{-}55)$$

As (6-55) is a quadratic equation there will be two solutions. For the coalesced doublet, one solution is a very broad line not experimentally observed and the other is the line we want. It has a limiting width at the fast exchange limit and is most easily obtained by writing

$$\Delta = C_0 + C_1\tau + C_2\tau^2 + \cdots. \quad (6\text{-}56)$$

Substituting Δ from (6-56) into (6-55) and equating powers of τ one obtains the result to lowest order in τ as

$$-(iC_0/\tau p_A p_B) - (1/\tau)[(1/p_B T_A) + (1/p_A T_B)] = 0 \quad (6\text{-}57)$$

or

$$C_0 = +i[(p_A/T_A) + (p_B/T_B)]. \quad (6\text{-}58)$$

The next higher order term in (6-55) will be

$$0 = C_0^2 - C_0\{\delta(p_A - p_B) + i[(1/T_A) + (1/T_B)]\} - C_1 i(1/p_A p_B) - p_A p_B \delta^2 - i\delta[(p_B/T_B) - (p_A/T_A)] - 1/T_A T_B. \quad (6\text{-}59)$$

Substituting for C_0 in (6-59) and letting $T_A = T_B$ (the general solution is given as a problem at the back of the chapter) we have

$$C_1 = ip_A^2 p_B^2 \delta^2. \quad (6\text{-}60)$$

Thus, to first order in τ we find the linewidth close to the fast exchange limit to be

$$d\omega = i[(p_A/T_A) + (p_B/T_B)] + ip_A^2 p_B^2 \delta^2 \tau. \quad (6\text{-}48)$$

The correction terms for $T_A \neq T_B$ in the term going like τ will be small.

(5) Coalescence Measurements

At the point where fine structure is first obscured, the so-called coalescence region Fig. 6-1, τ can be obtained quite simply. As mentioned above, absorption maxima exist where D [in the denominator of the right-hand side of (6-42)] goes to zero. The frequencies are obtained by solving the equation in Δ (6-55), which is of the form

$$\Delta^2 + \Delta B + C = 0,$$

with solution

$$\Delta = \tfrac{1}{2}\left[1 \pm \sqrt{B^2 - 4AC}\,\right]. \quad (6\text{-}61)$$

As the square root term passes from being real through zero to imaginary the lineshape changes from a doublet, through coalescence to a single line, respectively. Thus, at coalescence we see that

$$\sqrt{B^2 - 4AC} = 0. \quad (6\text{-}62)$$

Taking $p_A = p_B = \frac{1}{2}$ and $T_A = T_B$ we find on solving (6-62) that

$$\tau = 2/\delta. \quad (6\text{-}63)$$

However, this expression is of limited value since the shift has to be measured under conditions of slow exchange, i.e., usually at lower temperatures. Also, the $\omega_{0A} - \omega_{0B}$ shift might vary with temperature. Then, the shift at coalescence has to be obtained by extrapolating from measurements made at *several* lower temperatures. In that case, one might as well carry out a total lineshape analysis in the first place.

(6) Further Applications of Two Site Exchange

Another useful application involves the exchange of species between sites with very different relaxation times, for example, the exchange of a proton between free water and a site on a protein near a paramagnetic center. When the exchange rate lies on the NMR time scale one can calculate sticking times from the NMR lineshapes. Also, in principle, equilibrium constants can be extricated from the data.

Altogether, the two site case provides an instructive example in which one can examine analytically the NMR lineshape behavior of an exchanging system.

b. Kubo–Anderson–Sack Formulation[5–11]

An extension of the treatment for the NMR lineshape of an uncoupled two site system undergoing exchange is to apply it to many sites.

Consider a collection of half-spin chemical species (containing, for instance, hydrogen), each of which occupies one among an array of magnetically nonequivalent sites. The sites carry labels A, B, C, D which also describe the species which occupy them, and the spins at each site give rise to single lines in the NMR spectrum. To obtain the absorption for such a system we need

$$\text{Abs} = -\text{Im}\sum_N (S_N)\tilde{\rho}^{N}_{\alpha,\beta}\,, \tag{6-64}$$

where (S_N) stands for the concentration of spins at site N.†

Let the species exchange places among the different sites M according to the general scheme

$$\text{N} \rightleftarrows \text{M}_1\,, \tag{6-65}$$

so that for a species at site N, the exchange term in the density matrix equation is

$$E\tilde{\rho}^{\text{N}} = \sum_M (\tilde{\rho}^{\text{M}} - \tilde{\rho}^{\text{N}})/\tau_{\text{N}\to\text{M}}\,. \tag{6-66}$$

To simplify matters, a linewidth parameter $1/T_{\text{N}}$ is used to account for nuclear relaxation at site N and field inhomogeneities due to the instrument. In the coupled density matrix equations, for simplicity, the state labels can be left off $\tilde{\rho}^{\text{N}}_{\alpha,\beta}$ since they all have the same form

$$\left[i(\omega - \omega_{0\text{N}}) - (1/T_{\text{N}}) - \sum_{\text{M}} (1/\tau_{\text{N}\to\text{M}})\right]\tilde{\rho}^{\text{N}} + \sum_{\text{M}} (1/\tau_{\text{N}\to\text{M}})\tilde{\rho}^{\text{M}} = i\beta'/2, \tag{6-67}$$

and an $\dot{\tilde{\rho}}^{\text{N}}$ equation is written for every site N.

Although solution of this equation requires inverting the $[i\omega\mathcal{I} + \text{A}]$ coefficient matrix, there is one condition, that of slow exchange, when rate constants can be measured from the individual linewidths. At slow exchange the off diagonal elements of the coefficient matrix can be neglected. Then, the individual elements of $\tilde{\rho}^{\text{N}}$ are obtained from (6-67) with off diagonal elements neglected as

$$\tilde{\rho}^{N} = \frac{i\beta'}{2}\left[\frac{1}{i(\omega - \omega_{0\text{N}}) - (1/T_{\text{N}}) - \Sigma_{\text{M}}(1/\tau_{\text{N}\to\text{M}})}\right], \tag{6-68}$$

which implies a half-width at half-height of the line representing the

†We use henceforth in this chapter, for simplicity, $\tilde{\rho}$ for $\tilde{\rho}_s$.

species at site N of

$$d\omega_{\text{slow}} = (1/T_{\text{N}}) + \sum_{\text{M}} 1/\tau_{\text{N}\to\text{M}} \,. \tag{6-69}$$

The multisite treatment given here is equivalent to that of Kubo, Anderson, and Sack based on McConnel's equations.[5–11] We have shown how the result comes easily and simply from density matrix theory.

c. *Weak Coupling Approximation*

A typical Hamiltonian for a nuclear spin system appears as

$$\mathcal{H}_0 = \sum_i \omega_{0i} I_i^z + \sum_{i>j} J_{i,j} \mathbf{I}_i \cdot \mathbf{I}_j \,. \tag{6-70}$$

For the experimental situation, in which

$$\omega_{0i} - \omega_{0j} \gg J_{i,j} \,, \tag{6-71}$$

one may approximate the Hamiltonian term

$$J_{i,j} \mathbf{I}_i \cdot \mathbf{I}_j \tag{6-72}$$

as

$$J_{i,j} \mathbf{I}_i^z \cdot \mathbf{I}_j^z \,. \tag{6-73}$$

This is the so-called weak coupling approximation. Its effect is to allow us to simply calculate energies for the i spins as

$$I_i^z \left[\omega_{0i} + \sum_j J_{i,j} I_j^z \right]. \tag{6-74}$$

Each assignment of m_j then gives rise to a separate line. As an example of weak coupling consider a molecule made up of one group of two equivalent $\frac{1}{2}$ spins (called A) and another group of three equivalent $\frac{1}{2}$ spins (called B). The spin states of the two spin group (labeled 1 and 2) are

ϕ	m_z
$\alpha_1\alpha_2$	1
$\alpha_1\beta_2\beta_1\alpha_2$	0
$\beta_1\beta_2$	-1

(6-75)

and the spin states of the three spin group (labeled 3, 4, 5) are

ϕ			m_z
	$\alpha_3\alpha_4\alpha_5$		$\frac{3}{2}$
$\alpha\alpha\beta$	$\alpha\beta\alpha$	$\beta\alpha\alpha$	$\frac{1}{2}$
$\beta\beta\alpha$	$\beta\alpha\beta$	$\alpha\beta\beta$	$-\frac{1}{2}$
	$\beta\beta\beta$		$-\frac{3}{2}$.

(6-76)

The local field seen by either spin of the A group will then be

$$\omega_{0,\,\mathrm{loc}} = \omega_{0\mathrm{A}} + J\sum_{i=1}^{3} \langle \gamma_3\gamma_4\gamma_5 | I_i^z | \gamma_3\gamma_4\gamma_5 \rangle, \tag{6-77}$$

where γ_i can be either α or β. Substituting from (6-74) we then have

$$\begin{aligned}
\omega_{0,\,\mathrm{loc}} &= \omega_{0\mathrm{A}} + J\left(\tfrac{1}{2}+\tfrac{1}{2}+\tfrac{1}{2}\right), & m_z &= \tfrac{3}{2},\\
\omega_{0,\,\mathrm{loc}} &= \omega_{0\mathrm{A}} + J\left(\tfrac{1}{2}+\tfrac{1}{2}-\tfrac{1}{2}\right), & m_z &= \tfrac{1}{2},\\
\omega_{0,\,\mathrm{loc}} &= \omega_{0\mathrm{A}} + J\left(\tfrac{1}{2}-\tfrac{1}{2}-\tfrac{1}{2}\right), & m_z &= -\tfrac{1}{2},\\
\omega_{0,\,\mathrm{loc}} &= \omega_{0\mathrm{A}} + J\left(-\tfrac{1}{2}-\tfrac{1}{2}-\tfrac{1}{2}\right), & m_z &= -\tfrac{3}{2}.
\end{aligned} \tag{6-78}$$

The total absorption from the two spin (A) combination will be twice that for a single spin. The local field from B will split the single absorption of the "isolated spin" into four peaks (6-78) with amplitude ratios 1, 3, 3, 1. Thus, the total absorption of the molecule will appear as Fig. 6-2.

d. Equivalent Spins, Equivalence Factoring[12]

Let us consider the problem of a molecule with two magnetically equivalent protons each exchanging with a proton on another molecule. Each species gives rise to a single line at slow exchange.

$$\mathrm{AHH} + \mathrm{BH}^* \rightleftarrows \mathrm{AHH}^* + \mathrm{BH}. \tag{6-79}$$

In the past, the problem of two magnetically equivalent spins $I = \frac{1}{2}$ has been handled by treating the system as one of spin $I = 1$. We shall now show how this quasi-solution can be avoided and how the real density

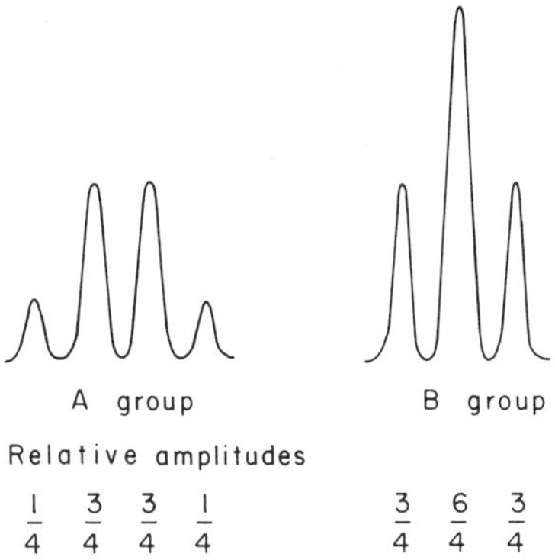

Fig. 6-2. Proton NMR of a weakly coupled two site five spin system AH_2BH_3 where two equivalent protons reside at site A and three at site B. The two sets of equivalent protons are weakly coupled.

matrix treatment leads to a 2 × 2 matrix to account for the two lines and their exchange behavior.

The two equivalent exchange processes, together with the state labels of the different protons, are written (the superscripts are labels of single protons) as

$$\underset{i \text{ of } \phi_i:}{} \quad \underset{ab}{\mathrm{AH^1H^2}} + \underset{c}{\mathrm{BH^3}} \overset{k}{\rightleftarrows} \underset{cb}{\mathrm{AH^3H^2}} + \underset{a}{\mathrm{BH^1}}, \tag{6-80}$$

$$\underset{i \text{ of } \phi_i:}{} \quad \underset{ab}{\mathrm{AH^1H^2}} + \underset{c}{\mathrm{BH^3}} \overset{k}{\rightleftarrows} \underset{ac}{\mathrm{AH^1H^3}} + \underset{b}{\mathrm{BH^2}}, \tag{6-81}$$

and the mean lifetimes for a proton in A and B are given by

$$1/\tau_{A2} = k(\mathrm{B}), \qquad 1/\tau_{\mathrm{B}} = k(\mathrm{A}). \tag{6-82}$$

According to Chapter V, the exchange terms in the density matrix equation for (6-80) are

$$(E\tilde{\rho}^{\mathrm{A}})_{\mathrm{ab,\,a'b'}} = (1/\tau_{\mathrm{A}})\left[\sum_{\mathrm{c}} \tilde{\rho}^{\mathrm{A}}_{\mathrm{cb,\,cb'}}\tilde{\rho}^{B}_{a,\,a'} + \sum_{c} \tilde{\rho}^{A}_{ac,\,a'c}\tilde{\rho}^{B}_{b,\,b'} - 2\tilde{\rho}^{\mathrm{A}}_{\mathrm{ab,\,a'b'}}\right], \tag{6-83}$$

$$\begin{aligned}(E\tilde{\rho}^{\mathrm{B}})_{\mathrm{b,\,b'}} &= (1/\tau_{\mathrm{B}})\left[\sum_{\mathrm{ab}} \tilde{\rho}^{\mathrm{A}}_{\mathrm{cb,c'b}}\tilde{\rho}^{\mathrm{B}}_{\mathrm{a,a}} + \sum_{\mathrm{ab}} \tilde{\rho}^{\mathrm{A}}_{\mathrm{ac,ac'}}\tilde{\rho}^{\mathrm{B}}_{\mathrm{b,b}} - 2\tilde{\rho}^{\mathrm{B}}_{\mathrm{b,b'}}\right] \\ &= (1/\tau_{\mathrm{B}})\left[\sum_{\mathrm{b}} \tilde{\rho}^{\mathrm{A}}_{\mathrm{cb,\,c'b}} + \sum_{\mathrm{a}} \tilde{\rho}^{\mathrm{A}}_{\mathrm{ac,\,ac'}} - 2\tilde{\rho}^{\mathrm{B}}_{\mathrm{b,\,b'}}\right],\end{aligned} \tag{6-84}$$

where we have used the relationship

$$\sum_{i} \tilde{\rho}_{i,\,i} = 1. \tag{6-85}$$

The low power density matrix equations which need to be solved are of the form

$$i\left[\tilde{\rho}^{\mathrm{A}}, \overline{\mathcal{H}}{}^{\mathrm{A}}_{0}\right] - (\tilde{\rho}^{\mathrm{A}}/T_{\mathrm{A}}) + E\tilde{\rho}^{\mathrm{A}} = -\left(\hbar\bar{\omega}_0\bar{\omega}_1/4kT\right)\sum_{s} I^{y}_{\mathrm{A}s}, \tag{6-86}$$

$$i\left[\tilde{\rho}^{\mathrm{B}}, \overline{\mathcal{H}}{}^{\mathrm{B}}_{0}\right] - (\tilde{\rho}^{\mathrm{B}}/T_{\mathrm{B}}) + E\tilde{\rho}^{\mathrm{B}} = -\left(\hbar\bar{\omega}_0\bar{\omega}_1/2kT\right)\sum I^{y}_{\mathrm{B}}. \tag{6-87}$$

For simplicity relaxation has been simulated only as $\tilde{\rho}/T$, the field inhomogeneity correction.

The basis functions and their labels are

ϕ_i^{A}	i	ϕ_i^{B}	i
$\alpha\alpha$	1	α	5
$\alpha\beta$	2	β	6,
$\beta\alpha$	3		
$\beta\beta$	4		

(6-88)

the shifts for A and B are ω_{0A} and ω_{0B} and $\not{\omega}_1$ is the rf field, and

$$\Delta\omega_A = \omega - \omega_{0A}\,. \tag{6-89}$$

To calculate the absorption,

$$\text{Abs} = -(A)\,\text{Im}\big(\tilde{\rho}^A_{1,2} + \tilde{\rho}^A_{1,3} + \tilde{\rho}^A_{2,4} + \tilde{\rho}^A_{3,4}\big) - (\text{B})\,\text{Im}\,\tilde{\rho}^B_{5,6}\,, \tag{6-90}$$

we need only matrix elements as they appear in (6-90). The required density matrix equations are then obtained from (6-86) and (6-87) as

$$\begin{aligned}
&\big[i\,\Delta\omega_A - (1/T_A)\big]\tilde{\rho}^A_{1,2} + (1/2\tau_A)\big(\tilde{\rho}^A_{3,4} + \tilde{\rho}^B_{5,6} - 3\tilde{\rho}^A_{1,2}\big) = i\beta'/4,\\
&\big[i\,\Delta\omega_A - (1/T_A)\big]\tilde{\rho}^A_{1,3} + (1/2\tau_A)\big(\tilde{\rho}^A_{2,4} + \tilde{\rho}^B_{5,6} - 3\tilde{\rho}^A_{1,3}\big) = i\beta'/4,\\
&\big[i\,\Delta\omega_A - (1/T_A)\big]\tilde{\rho}^A_{2,4} + (1/2\tau_A)\big(\tilde{\rho}^A_{1,3} + \tilde{\rho}^B_{5,6} - 3\tilde{\rho}^A_{2,4}\big) = i\beta'/4,\\
&\big[i\,\Delta\omega_A - (1/T_A)\big]\tilde{\rho}^A_{3,4} + (1/2\tau_A)\big(\tilde{\rho}^A_{1,2} + \tilde{\rho}^B_{5,6} - 3\tilde{\rho}^A_{3,4}\big) = i\beta'/4,\\
&\big[i\,\Delta\omega_B - (1/T_B)\big]\tilde{\rho}^B_{5,6} + (1/\tau_B)\big(\tilde{\rho}^A_{1,2} + \tilde{\rho}^A_{1,3} + \tilde{\rho}^A_{2,4} + \tilde{\rho}^A_{3,4} - 2\tilde{\rho}^B_{5,6}\big)\\
&\quad = i\beta'/2.
\end{aligned} \tag{6-91}$$

From (6-91) it is clear that each of the first four equations are permutations of each other and thus equivalent. To take advantage of this equivalency, we add the first four equations, define

$$\tilde{\rho}_I = \tilde{\rho}^A_{1,2} + \tilde{\rho}^A_{1,3} + \tilde{\rho}^A_{2,4} + \tilde{\rho}^A_{3,4}, \tag{6-92}$$

and write (6-91) as

$$\begin{aligned}
&\big[i\,\Delta\omega_{0A} - (1/T_A) - (1/\tau_A)\big]\tilde{\rho}^A_I + \tilde{\rho}^B_{5,6}(2/\tau_A) = i\beta',\\
&\big[i\,\Delta\omega_{0B} - (1/T_B) - (2/\tau_B)\big]\tilde{\rho}_{5,6} + \tilde{\rho}^A_I(1/\tau_B) = \tfrac{1}{2}i\beta'.
\end{aligned} \tag{6-93}$$

Notice that (6-93) are just two coupled equations for the two line spectrum. For the exchange of two protons, each in a different site, we would need the equations given in (6-39)—note the difference.

Next, let us exhibit the spectrum of the system described by (6-79) under conditions of very slow exchange (see Fig. 6-3). The amplitudes are relative to a single spin $\frac{1}{2}$. If we associate with each peak a spin $\frac{1}{2}$ absorbtion, ρ_s, we then have the relation

$$\tilde{\rho}^A_I = 2\tilde{\rho}^A_{Is}\,, \tag{6-94}$$

$$\tilde{\rho}^B_{5,6} = \tilde{\rho}^A_{5,6_s}\,. \tag{6-95}$$

Substituting (6-94) and (6-95) into (6-93) one obtains

$$\begin{aligned}
&\big[\Delta\omega_A - (1/T_A) - (1/\tau_A)\big]\tilde{\rho}^A_{Is} + \tilde{\rho}^B_{5,6_s}\;1/\tau_A = \tfrac{1}{2}i\beta',\\
&\big[i\,\Delta\omega_B - (1/T_B) - (2/\tau_B)\big]\tilde{\rho}_{5,6s} + \tilde{\rho}^A_{Is}(2/\tau_B) = \tfrac{1}{2}i\beta'.
\end{aligned} \tag{6-96}$$

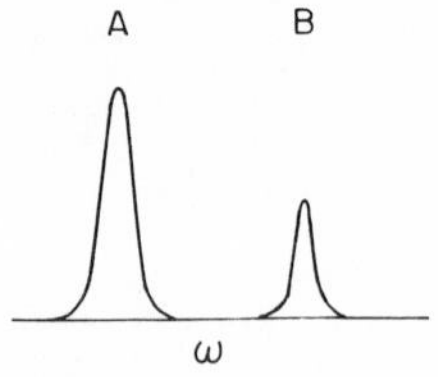

Fig. 6-3. The NMR spectrum of system (6-80), (6-81).

Note that (6-96) has the appearance of two spin $\frac{1}{2}$ sites (note the $i\beta/2$ term on the right—the signature of a $\frac{1}{2}$ spin site) exchanging with each other with specific rates

$$1/\tau_A \quad \text{and} \quad 2/\tau_B , \tag{6-97}$$

which satisfies the Kubo–Anderson–Sack relations (6-60).

The generalization of (6-94) for n equivalent sites

$$AH_n + BH^* \rightleftarrows AH_{n-1}H^* + BH \tag{6-98}$$

is given as

$$\begin{aligned} \left[i\,\Delta\omega_A - (1/T_A) - (1/\tau_{As})\right]\tilde{\rho}^{A}_{Is} + \rho_{5,6s}/\tau_{As} &= \tfrac{1}{2}i\beta', \\ \left[i\,\Delta\omega_B - (1/T_B) - (1/\tau_{Bs})\right]\tilde{\rho}^{B}_{5,6s} + \tilde{\rho}^{A}_{Is}/\tau_{Bs} &= \tfrac{1}{2}i\beta', \end{aligned} \tag{6-99}$$

where

$$\text{Abs} = -n\left[(A)\text{Im}\,\tilde{\rho}^{A}_{Is}\right] - (B)\text{Im}\,\tilde{\rho}^{B}_{5,6s} , \tag{6-100}$$

$$1/\tau_{As} = 1/\tau_A , \qquad 1/\tau_{Bs} = n/\tau_B . \tag{6-101}$$

The simple interpretation is that the single spin on B has n chances of hopping onto A, where the rate $1/\tau_B$ is for a single event.

Consider a more complex equivalency problem, the exchange process

$$\begin{aligned} AH^1H^2BH^3 + GH^4 &\rightleftarrows AH^4H^2BH^3 + GH^1, \\ AH^1H^2BH^3 + GH^4 &\rightleftarrows AH^1H^4BH^3 + GH^2, \end{aligned} \tag{6-102}$$

where A, B, and G denote magnetically nonequivalent sites and the superscript numbers are labels of single hydrogens. Under conditions of slow exchange the spectrum will appear as Fig. 6-4, where the amplitudes are given relative to a single spin $\frac{1}{2}$. The Hamiltonians in the rotating coordinate system are

$$\overline{\mathcal{H}}_0^{AB} = (\omega_{0A} - \omega)(I_1^z + I_2^z) + (\omega_{0B} - \omega)I_3^z + J_{AB}I_3^z \cdot (I_1^z + I_2^z) \tag{6-103}$$

and

$$\overline{\mathcal{H}}_0^{G} = (\omega_{0G} - \omega)I_4^z . \tag{6-104}$$

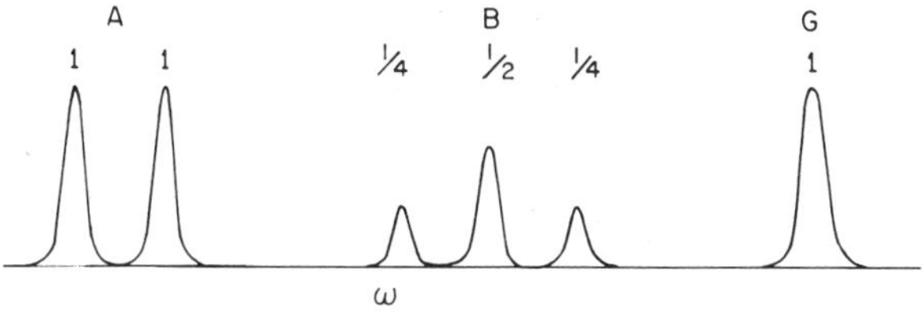

Fig. 6-4. The proton NMR spectrum of the weakly coupled system AH_2BHGH where A, B, and G are site labels. Only observable coupling is $J(AH_2, BH)$.

The product wave functions are given as

ϕ_i	i	ϕ_i	i	ϕ_i	i
$\alpha_1\alpha_2\alpha_3$	1	$\beta_1\beta_2\alpha_3$	5	α	9
$\alpha\alpha\beta$	2	$\beta\alpha\beta$	6	β	10
$\alpha\beta\alpha$	3	$\alpha\beta\beta$	7		
$\beta\alpha\alpha$	4	$\beta\beta\beta$	8		

For the low power spectrum we need the $\Delta m = 1$ transitions

$$\left\{\begin{matrix} 1\text{–}3 \\ 1\text{–}4 \\ \hline 3\text{–}5 \\ 4\text{–}5 \end{matrix}\right\} \qquad \left\{\begin{matrix} 2\text{–}7 \\ 2\text{–}6 \\ \hline 7\text{–}8 \\ 6\text{–}8 \end{matrix}\right\}$$

$$\{1\text{–}2\} \qquad \left\{\begin{matrix} 3\text{–}7 \\ 4\text{–}6 \end{matrix}\right\} \qquad \{5\text{–}8\}$$

$$\{9\text{–}10\}$$

$$\left\{\begin{matrix} 4\text{–}7 \\ 3\text{–}6 \\ 2\text{–}5 \end{matrix}\right\} \quad \text{combination transitions}$$

Now, we must distinguish a critical difference between using the weak coupling approximation and using the full Hamiltonian. For weak coupling all terms in each bracket are equivalent and the combination terms are dropped. The equivalent density matrix elements in brackets are then represented as

$$\tilde{\rho}_{\rm I} = \tilde{\rho}^{\rm AB}_{1,3} + \rho^{\rm AB}_{1,4} + \tilde{\rho}^{\rm AB}_{3,5} + \tilde{\rho}^{\rm AB}_{4,5}, \tag{6-105}$$

$$\tilde{\rho}_{\rm II} = \tilde{\rho}^{\rm AB}_{2,7} + \tilde{\rho}^{\rm AB}_{2,6} + \tilde{\rho}^{\rm AB}_{7,8} + \tilde{\rho}^{\rm AB}_{6,8}, \tag{6-106}$$

$$\tilde{\rho}_{\rm III} = \tilde{\rho}^{\rm AB}_{1,2}, \tag{6-107}$$

$$\tilde{\rho}_{\rm IV} = \tilde{\rho}^{\rm AB}_{3,7} + \tilde{\rho}^{\rm AB}_{4,6}, \tag{6-108}$$

$$\tilde{\rho}_{\rm V} = \tilde{\rho}^{\rm AB}_{5,8}, \tag{6-109}$$

$$\tilde{\rho}_{\rm VI} \doteq \tilde{\rho}^{\rm G}_{9,10}. \tag{6-110}$$

For weak coupling and slow exchange there is a one-to-one correspondence between the $\tilde{\rho}_{\mathrm{I}}$, $\tilde{\rho}_{\mathrm{II}}$, etc., density matrix elements, and the absorption spectrum illustrated in Fig. 6-4, i.e., the absorption is given as

$$\mathrm{Abs} = -(\mathrm{AB})\,\mathrm{Im}(\tilde{\rho}_{\mathrm{I}} + \tilde{\rho}_{\mathrm{II}} + \tilde{\rho}_{\mathrm{III}} + \tilde{\rho}_{\mathrm{IV}} + \tilde{\rho}_{\mathrm{V}}) - (\mathrm{G})\mathrm{Im}\,\tilde{\rho}_{\mathrm{VI}}\,. \quad (6\text{-}111)$$

The numerical relationships between the true ρ_{I}'s, etc., and the pseudo $\rho_{\mathrm{I}s}$'s, etc., are given as[§]

$$\tilde{\rho}_{\mathrm{I}} = \tilde{\rho}_{\mathrm{I}s}\,, \qquad \tilde{\rho}_{\mathrm{II}} = \tilde{\rho}_{\mathrm{II}s}\,, \qquad \tilde{\rho}_{\mathrm{III}} = \tfrac{1}{4}\tilde{\rho}_{\mathrm{III}s}\,,$$
$$\tilde{\rho}_{\mathrm{IV}} = \tfrac{1}{2}\tilde{\rho}_{\mathrm{IV}s}\,, \qquad \tilde{\rho}_{\mathrm{V}} = \tfrac{1}{4}\tilde{\rho}_{\mathrm{V}s}\,, \qquad \tilde{\rho}_{\mathrm{VI}} = \tilde{\rho}_{\mathrm{VI}s}\,. \quad (6\text{-}112)$$

To obtain the density matrix equations for $\tilde{\rho}_{\mathrm{I}}$ (and also $\tilde{\rho}_{\mathrm{I}s}$), one must first obtain the density matrix equations for $\tilde{\rho}^{\mathrm{AB}}_{1,3}$, $\tilde{\rho}^{\mathrm{AB}}_{1,4}$, etc. For $\tilde{\rho}^{\mathrm{AB}}_{1,3}$ we have the exchange contribution

$$(E\tilde{\rho}^{\mathrm{AB}})_{1,3} = (1/\tau_{\mathrm{AB}})\Big[\sum_c \tilde{\rho}^{\mathrm{AB}}_{c\alpha\alpha,\,c\beta\alpha}\tilde{\rho}^{\mathrm{G}}_{\alpha,\,\alpha} - \tilde{\rho}^{\mathrm{AB}}_{\alpha\alpha\alpha,\,\alpha\beta\alpha} + \tilde{\rho}^{\mathrm{AB}}_{\alpha c\alpha,\,\alpha c\alpha}\tilde{\rho}^{\mathrm{G}}_{\alpha,\,\beta} - \tilde{\rho}^{\mathrm{AB}}_{\alpha\alpha\alpha,\,\alpha\beta\alpha}\Big]$$
$$= (1/\tau_{\mathrm{AB}})\Big[\tfrac{1}{2}\tilde{\rho}^{\mathrm{AB}}_{4,5} + \tfrac{1}{4}\tilde{\rho}^{\mathrm{AB}}_{9,10} - \tfrac{3}{2}\tilde{\rho}^{\mathrm{AB}}_{1,3}\Big], \quad (6\text{-}113)$$

where we have linearized the equation using the Selective Neglect of Bilinear terms (SNOB) approximation. Adding up the equivalent transitions in $\tilde{\rho}_{\mathrm{I}}$ (6-105)

$$(i\,\Delta\omega_{\mathrm{I}} - 1/T)\tilde{\rho}_{\mathrm{I}} + (1/\tau_{\mathrm{AB}})(\tilde{\rho}_{\mathrm{VI}} - \tilde{\rho}_{\mathrm{I}}) = \tfrac{1}{2}i\beta', \quad (6\text{-}114)$$

where $1/T$ is the field inhomogeneity parameter. In similar fashion, one obtains

$$(i\,\Delta\omega_{\mathrm{II}} - 1/T)\tilde{\rho}_{\mathrm{II}} + (1/\tau_{\mathrm{AB}})(\tilde{\rho}_{\mathrm{VI}} - \tilde{\rho}_{\mathrm{II}}) = \tfrac{1}{2}i\beta', \quad (6\text{-}115)$$
$$(i\,\Delta\omega_{\mathrm{III}} - 1/T)\tilde{\rho}_{\mathrm{III}} + (1/\tau_{\mathrm{AB}})\big(\tfrac{1}{2}\tilde{\rho}_{\mathrm{IV}} - \tilde{\rho}_{\mathrm{III}}\big) = \tfrac{1}{8}i\beta', \quad (6\text{-}116)$$
$$(i\,\Delta\omega_{\mathrm{IV}} - 1/T)\tilde{\rho}_{\mathrm{IV}} + (1/\tau_{\mathrm{AB}})(\tilde{\rho}_{\mathrm{III}} + \tilde{\rho}_{\mathrm{V}} - \tilde{\rho}_{\mathrm{IV}}) = \tfrac{1}{4}i\beta', \quad (6\text{-}117)$$
$$(i\,\Delta\omega_{\mathrm{V}} - 1/T)\tilde{\rho}_{\mathrm{V}} + (1/\tau_{\mathrm{AB}})(\tilde{\rho}_{\mathrm{IV}} - \tilde{\rho}_{\mathrm{V}}) = \tfrac{1}{8}i\beta', \quad (6\text{-}118)$$
$$(i\,\Delta\omega_{\mathrm{VI}} - 1/T)\tilde{\rho}_{\mathrm{VI}} + (1/\tau_{\mathrm{G}})(\tilde{\rho}_{\mathrm{I}} + \tilde{\rho}_{\mathrm{II}} - 2\tilde{\rho}_{\mathrm{VI}}) = \tfrac{1}{2}i\beta'. \quad (6\text{-}119)$$

[§]The results given in (6-112) can be obtained by noting, for example, that

$$\tilde{\rho}_{1,3} = \langle\alpha\alpha\alpha|\tilde{\rho}^{\mathrm{AB}}|\alpha\beta\alpha\rangle. \quad (6\text{F-}1)$$

Then dimensionally we can write

$$\tilde{\rho}_{1,3} = \langle\alpha\alpha\|\alpha\alpha\rangle\langle\alpha\|\,\beta\rangle.$$
$$= \frac{1}{2^2}\rho_s = \frac{1}{4}\rho_s \quad (6\text{F-}2)$$

as $\langle\alpha\alpha\|\alpha\alpha\rangle$is one of four possible diagonal states. Thus,

$$\tilde{\rho}_{\mathrm{IV}} = \tilde{\rho}^{\mathrm{AB}}_{3,7} + \tilde{\rho}^{\mathrm{AB}}_{3,6} = \tfrac{1}{4}\tilde{\rho}_{\mathrm{IV}_s} + \tfrac{1}{4}\tilde{\rho}_{\mathrm{IV}_s} = \tfrac{1}{2}\tilde{\rho}_{\mathrm{IV}_s} \quad (6\text{F-}3)$$

The frequencies $\Delta\omega$, etc., are related to the shifts and coupling constants as follows:

$$\Delta\omega_{\rm I} = \omega - \omega_{0\rm A} + \tfrac{1}{2}J_{\rm AB}\,, \qquad \Delta\omega_{\rm II} = \omega - \omega_{0\rm A} - \tfrac{1}{2}J_{\rm AB}\,,$$

$$\Delta\omega_{\rm III} = \omega - \omega_{0\rm B} + J_{\rm AB}\,, \qquad \Delta\omega_{\rm IV} = \omega - \omega_{0\rm B}\,,$$

$$\Delta\omega_{\rm V} = \omega - \omega_{0\rm B} - J_{\rm AB}\,, \qquad \Delta\omega_{\rm VI} = \omega - \omega_{0\rm G}\,. \tag{6-120}$$

Substituting the $\tilde{\rho}_{{\rm I}s}$ terms for the summed terms, $\tilde{\rho}_{\rm I}$, we see that

$$(i\,\Delta\omega_{\rm I} - 1/T)\tilde{\rho}_{{\rm I}s} + (1/\tau_{\rm AB})(\tilde{\rho}_{{\rm VI}s} - \tilde{\rho}_{{\rm I}s}) = \tfrac{1}{2}i\beta', \tag{6-121}$$

$$(i\,\Delta\omega_{\rm II} - 1/T)\tilde{\rho}_{{\rm II}s} + (1/\tau_{\rm AB})(\tilde{\rho}_{{\rm VI}s} - \tilde{\rho}_{{\rm II}s}) = \tfrac{1}{2}i\beta', \tag{6-122}$$

$$(i\,\Delta\omega_{\rm III} - 1/T)\tilde{\rho}_{{\rm III}s} + (1/\tau_{\rm AB})(\tilde{\rho}_{{\rm IV}s} - \tilde{\rho}_{{\rm III}s}) = \tfrac{1}{2}i\beta', \tag{6-123}$$

$$i(\Delta\omega_{\rm IV} - 1/T)\tilde{\rho}_{{\rm IV}s} + (1/2\tau_{\rm AB})(\tilde{\rho}_{{\rm III}s} + \tilde{\rho}_{{\rm V}s} - 2\tilde{\rho}_{{\rm IV}s}) = \tfrac{1}{2}i\beta', \tag{6-124}$$

$$i(\Delta\omega_{\rm V} - 1/T)\tilde{\rho}_{{\rm V}s} + (1/\tau_{\rm AB})(\tilde{\rho}_{{\rm IV}s} - \tilde{\rho}_{{\rm V}s}) = \tfrac{1}{2}i\beta', \tag{6-125}$$

$$i(\Delta\omega_{\rm VI} - 1/T)\tilde{\rho}_{{\rm VI}s} + (1/\tau_{\rm G})(\tilde{\rho}_{{\rm I}s} + \tilde{\rho}_{{\rm II}s} - 2\tilde{\rho}_{{\rm VI}s}) = \tfrac{1}{2}i\beta'. \tag{6-126}$$

Noting the signature of a spin $\frac{1}{2}$, $i\beta'\frac{1}{2}$, on the right-hand side of each of the last six equations, we see we have reduced the problem (in the low power weak coupling limit) to that of a set of six nonequivalent $\frac{1}{2}$ spin resonances. One also sees that the Kubo–Anderson–Sack relations are formally satisfied. However, only by working out the problem in detail, has it been possible with confidence to identify the Kubo–Anderson–Sack τ's in terms of those kinetic parameters which describe the proposed mechanism.

e. Exchange Averaging of the Coupling in Weakly Coupled Systems, Examples

(1) Exchange in Hydrogen Fluoride[13]

It has already been shown how the NMR lineshape of hydrogen fluoride depends on the HF bond exchange rate in the process

$$\underset{\rm ab}{\rm H^*F^*} + \underset{\rm cd}{\rm HF} \overset{k}{\rightleftarrows} \underset{\rm ac}{\rm H^*F} + \underset{\rm bd}{\rm HF^*}. \qquad (i \text{ of } \phi_i) \tag{6-127}$$

The NMR spectrum consists of two doublets for the proton and fluorine resonances, respectively, the spitting being due to J(H, F) of 615 Hz. When fluorine jumps between hydrogens in different spin states, this coupling becomes averaged out and in the fast exchange limit single lines are obtained for the proton and fluorine resonances.

For this system we list the spin states of the HF molecule $\alpha_H\alpha_F$, $\alpha\beta$, $\beta\alpha$, and $\beta\beta$ and label them 1, 2, 3, and 4, respectively. As described in Chapter V, $E\tilde{\rho}$ is given by

$$[E\tilde{\rho}^{HG}]_{ab,\,a'b'} = k(\mathrm{HF})\sum_d \tilde{\rho}^{HF}_{ad,\,a'd}\sum_c \tilde{\rho}^{HF}_{cb,\,cb'} - k(\mathrm{HF})\tilde{\rho}^{HF}_{ab,\,a'b'}\,. \tag{6-128}$$

Linewidth parameters which include effects due to field inhomogeneity are $1/T_H$ and $1/T_F$, shifts are ω_{0H} and ω_{0F}, differences are defined as

$$\Delta\omega_H = \omega - \omega_{0H}\,, \qquad \Delta\omega_F = \omega - \omega_{0F}\,, \tag{6-129}$$

and J is the H, F coupling constant. Equation (6-131) on page 92 displays the coupled equations in matrix form where the proportionality constants β' and β'' are given as

$$\beta' = \hbar\omega_{0F}\omega_{1F}/2kT, \tag{6-130a}$$

$$\beta' = \hbar\omega_{0H}\omega_{1H}/2kT. \tag{6-130b}$$

Since we are dealing with a weakly coupled system, the proton and fluorine transitions do not mix. The exchange process only averages the coupling constant. Thus, we can solve for the proton and fluorine resonances separately. Each of the submatrices (blocked out) looks like a matrix for an uncoupled two site problem, in the case of fluorine, with "shifts" $\Delta\omega_F \pm J/2$ and rate constant $k/2$. The latter factor of $\frac{1}{2}$ accounts for the fact that in each exchange there is a 50% probability that when fluorine exchanges from one HF to another the spin state of the hydrogen in the new HF differs from the first; only exchanges of this latter type average the coupling constant. This is an important point. Thus, we see that HF exchange can be treated as two independent, two site Kubo–Anderson–Sack cases. Notice that the same kinetic information can be obtained from either the proton or the fluorine resonance.

Suppose now that the system contains HF and fluoride ions and the exchange process includes the step

$$\underset{i \text{ of } \phi_i:}{} \quad \underset{e}{\overset{*}{F}{}^-} + \underset{ab}{HF} \overset{k'}{\rightleftarrows} \underset{b}{F^-} + \underset{ae}{HF^*} \tag{6-132}$$

in addition to (6-127). The problem now includes the coupled density matrix equations for the elements of HF and F^-. We label the states of F^-, α, and β, 5, and 6, respectively, and define

$$\Delta\omega_{F^-} = \omega - \omega_{0F^-}\,,$$

where ω_{0F^-} is the shift for fluoride anion. This time the $E\tilde{\rho}$ contributions refer to reactions (6-127) and (6-132)

$$\left[E\tilde{\rho}^{HF}\right]_{ab,\,a'b'} = k(\mathrm{HF})\left[\sum_{d}\tilde{\rho}^{HF}_{ad,\,a'd}\sum_{c}\tilde{\rho}^{HF}_{cb,\,cb'} - \tilde{\rho}^{HF}_{ab,\,a'b'}\right]$$

$$+\,k'(\mathrm{F}^-)\left[\sum_{e}\tilde{\rho}^{HF}_{ae,\,a'e}\tilde{\rho}^{F^-}_{b,\,b'} - \tilde{\rho}^{HF}_{ab,\,a'b'}\right], \qquad (6\text{-}133)$$

$$\left[\tilde{\rho}^{F^-}\right]_{e,\,e'} = k'(\mathrm{HF})\left[\sum_{a}\tilde{\rho}^{HF}_{ae,\,ae'} - \tilde{\rho}^{F^-}_{e,\,e'}\right]. \qquad (6\text{-}134)$$

The $\langle 1|E\tilde{\rho}^{HF}|2\rangle$ linearized element is given as

$$\left[E\tilde{\rho}^{HF}\right]_{1,2} = \tfrac{1}{2}k(\mathrm{HF})\left[\tilde{\rho}^{HF}_{3,4} - \tilde{\rho}^{HF}_{1,2}\right]$$

$$+\,k'(\mathrm{F}^-)\left[\tfrac{1}{2}\tilde{\rho}^{F^-}_{5,6} - \tilde{\rho}^{HF}_{1,2}\right]. \qquad (6\text{-}135)$$

The resulting density matrix equations for exchange in both reactions are given by (6-136) on page 92. Note that the proton and fluorine resonances can still be calculated separately.

(2) Bond Exchange in Trimethylthallium[14]

Trimethylthallium contains two isotopic species, one with ^{203}Tl and the other ^{205}Tl, each spin $\frac{1}{2}$ and of isotopic abundances 29.1% and 70.9%, respectively. The proton resonance at low temperatures consists of four lines, one pair for each isotopic species separated by the coupling constants of 250.7 and 248.3 Hz, respectively, for $J(^{205}\mathrm{Tl,H})$ and $J(^{203}\mathrm{TlH})$. At elevated temperatures these resonances coalesce due to intermolecular methyl thallium exchange

(6-137)

Kinetic studies due to Evans imply that exchange is second order in trimethylthallium and takes place via the doubly bridged transition state **1**.

$$
\begin{bmatrix}
\begin{pmatrix} i\,\Delta\omega_{\mathrm{F}} - (iJ/2) \\ -(1/T_{\mathrm{F}}) - k(\mathrm{HF})/2 \end{pmatrix} & k(\mathrm{HF})/2 & 0 & 0 \\
k(\mathrm{HF})/2 & \begin{pmatrix} i\,\Delta\omega_{\mathrm{F}} + (iJ/2) \\ -(1/T_{\mathrm{F}}) - k(\mathrm{HF})/2 \end{pmatrix} & 0 & 0 \\
0 & 0 & \begin{pmatrix} i\,\Delta\omega_{\mathrm{H}} - (iJ/2) \\ -(1/T_{\mathrm{H}}) - k(\mathrm{HF})/2 \end{pmatrix} & k(\mathrm{HF})/2 \\
0 & 0 & k(\mathrm{HF})/2 & \begin{pmatrix} i\,\Delta\omega_{\mathrm{H}} + (iJ/2) \\ -(1/T_{\mathrm{H}}) - k(\mathrm{HF})/2 \end{pmatrix}
\end{bmatrix}
\begin{bmatrix} \tilde{\rho}^{\mathrm{HF}}_{1,2} \\ \tilde{\rho}^{\mathrm{HF}}_{3,4} \\ \tilde{\rho}^{\mathrm{HF}}_{1,3} \\ \tilde{\rho}^{\mathrm{HF}}_{2,4} \end{bmatrix}
=
\begin{bmatrix} i\beta'/4 \\ i\beta'/4 \\ i\beta''/4 \\ i\beta''/4 \end{bmatrix}
\tag{6-131}
$$

$$
\begin{bmatrix}
\begin{pmatrix} i(\Delta\omega_{\mathrm{F}} - J/2) \\ -(1/T_{\mathrm{F}}) - [k(\mathrm{HF})/2] - k'(\mathrm{F}^-) \end{pmatrix} & k(\mathrm{HF})/2 & k'(\mathrm{F}^-)/2 & 0 & 0 \\
k(\mathrm{HF})/2 & \begin{pmatrix} i(\Delta\omega_{\mathrm{F}} - J/2) \\ -(1/T_{\mathrm{F}}) - [k(\mathrm{HF})/2] - k'(\mathrm{F}^-) \end{pmatrix} & k'(\mathrm{F}^-)/2 & 0 & 0 \\
k'(\mathrm{F}^-) & k'(\mathrm{F}^-) & \begin{pmatrix} i\,\Delta\omega_{\mathrm{F}^-} - (1/T_{\mathrm{F}}) - k'(\mathrm{F}^-) \end{pmatrix} & 0 & 0 \\
0 & 0 & 0 & \begin{pmatrix} i(\Delta\omega_{\mathrm{H}} - J/2) \\ -(1/T_{\mathrm{H}}) - [k(\mathrm{HF})/2] - k'(\mathrm{F}^-)/2 \end{pmatrix} & k(\mathrm{HF})/2 + k'(\mathrm{F}^-)/2 \\
0 & 0 & 0 & [k(\mathrm{HF})/2] + k'(\mathrm{F}^-)/2 & \begin{pmatrix} i(\Delta\omega_{\mathrm{H}} + J/2) \\ -(1/T_{\mathrm{H}}) - [k(\mathrm{HF})/2] - k'(\mathrm{F}^-)/2 \end{pmatrix}
\end{bmatrix}
\begin{bmatrix} \tilde{\rho}^{\mathrm{HF}}_{1,2} \\ \tilde{\rho}^{\mathrm{HF}}_{3,4} \\ \tilde{\rho}^{\mathrm{F}^-}_{5,6} \\ \tilde{\rho}^{\mathrm{HF}}_{1,3} \\ \tilde{\rho}^{\mathrm{HF}}_{2,4} \end{bmatrix}
= i
\begin{bmatrix} \beta'/4 \\ \beta'/4 \\ \beta'/2 \\ \beta''/4 \\ \beta''/4 \end{bmatrix}
\tag{6-136}
$$

This problem

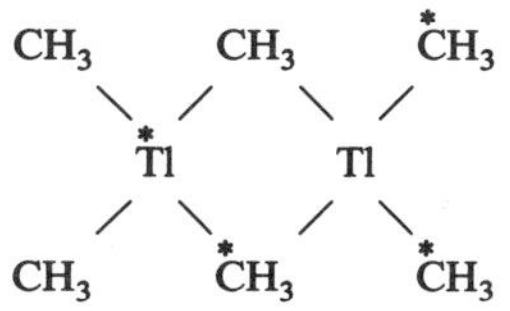

1

has previously been treated with the Kubo–Anderson–Sack method and simulated as an exchange among uncoupled magnetically nonequivalent sites. We now describe a simplified density matrix treatment which takes account of the proposed mechanism of exchange. The procedure which follows correctly simulates only the proton part of the trimethylthallium NMR spectrum.

First, we simulate trimethylthallium as "TlH" where H represents all nine methyl hydrogens. The different exchange processes to be considered are

$$\mathrm{Tl^*H^* + TlH \overset{k'}{\rightleftarrows} Tl^*H + TlH^*}$$

i of ϕ_i:	ab	cd	ac	bd,	(6-138)
	203	203	203	203,	(6-138a)
	203	205	203	205,	(6-138b)
	205	205	205	205,	(6-138c)

where the numbers are the thallium isotopic masses for different species involved and ab, cd are product functions for "TlH". We assume isotope effects are too small to distinguish among the rate constants for (6-138a)–(6-138c) using NMR methods.

Mean lifetimes for the different species in the three reactions (6-138a)–(6-138c) are

$$1/\tau_{203,\,\mathrm{a}} = k'(^{203}\mathrm{TlH}) = 0.291k'(\mathrm{TlH}), \tag{6-139}$$

$$1/\tau_{203,\,\mathrm{b}} = 0.709k'(\mathrm{TlH}), \tag{6-140}$$

$$1/\tau_{205,\,\mathrm{b}} = 0.291k'(\mathrm{TlH}), \tag{6-141}$$

$$1/\tau_{205,\,\mathrm{c}} = 0.791k'(\mathrm{TlH}), \tag{6-142}$$

where $\tau_{205\mathrm{c}}$ means preexchange lifetime of ^{205}TlH with respect to process (6-138c).

We list the four basis functions of ^{203}TlH and ^{205}TlH to be

^{203}TlH		^{205}TlH	
ϕ_i	i	ϕ_i	i
$\alpha\alpha$	1	$\alpha\alpha$	5
$\alpha\beta$	2	$\alpha\beta$	6
$\beta\alpha$	3	$\beta\alpha$	7
$\beta\beta$	4	$\beta\beta$	8

and the density matrix elements needed to calculate the proton absorption are

$$\tilde{\rho}_{1,2}^{203}, \tilde{\rho}_{3,4}^{203}, \tilde{\rho}_{5,6}^{205}, \tilde{\rho}_{7,8}^{205}, \tag{6-143}$$

where the superscripts serve as species labels. The absorption is then given as

$$\text{Abs} = -0.709(9c)\,\text{Im}\left(\tilde{\rho}_{1,2}^{203} + \tilde{\rho}_{3,4}^{203}\right) - 0.291(9c)\,\text{Im}\left(\tilde{\rho}_{5,6}^{205} + \tilde{\rho}_{7,8}^{205}\right), \tag{6-144}$$

where (c) is the total concentration of trimethylthallium and the factor of nine converts to concentrations of hydrogen.

The $E\rho$ contributions to the density matrix equations are evaluated according to the procedures of Chapter V. For instance, the result for $[E\tilde{\rho}^{203}]_{1,2}$ is given as

$$\left[E\tilde{\rho}^{203}\right]_{1,2} = \frac{1}{2\tau_{203,a}}\left[\tilde{\rho}_{3,4}^{203} - \tilde{\rho}_{1,2}^{203}\right] + \frac{1}{\tau_{203,b}}\left[\frac{\tilde{\rho}_{5,6}^{205} + \tilde{\rho}_{7,8}^{205}}{2} - \tilde{\rho}_{1,2}^{203}\right], \tag{6-145}$$

which on substitution of the definitions in (6-108)–(6-111) becomes

$$\left[E\tilde{\rho}^{203}\right]_{1,2} = k'(\text{TlH})\left[0.146\tilde{\rho}_{3,4}^{203} + 0.355\tilde{\rho}_{5,6}^{205} + 0.355\tilde{\rho}_{7,8}^{205} - 0.855\tilde{\rho}_{1,2}^{203}\right]. \tag{6-146}$$

Note, as trimethylthallium is a weakly coupled system, off diagonal elements of the spin Hamiltonian can be neglected. Since the thallium and proton transitions do not mix, the proton absorption is solved for separately.

Finally, to make the connection between the simulated (6-138) and proposed mechanism (6-137) it is necessary to replace

$$1/\tau_{\text{TlH}} \quad \text{by} \quad 3/\tau_{\text{TlMe}_3} \tag{6-147}$$

and

$$k'(\text{TlH}) \quad \text{by} \quad 3k(\text{TlMe}_3). \tag{6-148}$$

The coupled density matrix equations, in matrix forms, obtained by the above procedure, are given as

$$\begin{bmatrix} \begin{pmatrix} i\,\Delta\omega_H - (i\,J_{203}/2) - (1/T) \\ -2.563k(c) \end{pmatrix} & 1.065k(c) & 0.438k(c) & 1.065k(c) \\ 0.438k(c) & \begin{pmatrix} i\,\Delta\omega_H - (i\,J_{205}/2) \\ -1.941k(c) - 1/T \end{pmatrix} & 0.438k(c) & 1.065k(c) \\ 0.438k(c) & 1.065k(c) & \begin{pmatrix} i\,\Delta\omega_H + (i\,J_{203}/2) \\ -2.653k(c) - 1/T \end{pmatrix} & 1.065k(c) \\ 0.438k(c) & 1.065k(c) & 0.438k(c) & \begin{pmatrix} i\,\Delta\omega_H + (i\,J_{205}/2) \\ -1.941k(c) - 1/T \end{pmatrix} \end{bmatrix} \begin{bmatrix} \tilde{\rho}^{203}_{1,2} \\ \tilde{\rho}^{205}_{5,6} \\ \tilde{\rho}^{203}_{3,4} \\ \tilde{\rho}^{205}_{7,8} \end{bmatrix} = \frac{i\beta'}{4} \begin{bmatrix} 1 \\ 1 \\ 1 \\ 1 \end{bmatrix} \tag{6-149}$$

where

$$\Delta\omega_H = \omega - \omega_{0H}\,, \tag{6-150}$$

J_{203} and J_{205} stand for the two thallium proton coupling constants, and (c) means total concentration of trimethylthallium.

We could have simulated the protons of each methyl group as a single spin $\frac{1}{2}$. The exchange to be simulated would involve two "H's" between two four spin molecules. All spin states and all $\Delta m = 1$ transitions of such a system would have to be taken into account! A great deal of equivalence factoring would give the same result as in Eq. (6-149).

Actually, the adjacent resonances due to $\tilde{\rho}^{203}_{1,2}$ and $\tilde{\rho}^{205}_{5,6}$ (or to the pair $\tilde{\rho}^{203}_{3,4}$ and $\tilde{\rho}^{205}_{7,8}$) are so close that they become averaged even at low temperature when exchange is slow. Because of this, the proton resonance for trimethylthallium has been previously simulated as a two site exchanging system.[19]

(3) Dimethylcadmium[15, 16]

Dimethylcadmium contains stable cadmium isotopes of masses 111, 113, 106, 108, 110, 112, and 114. Of these the first two, 12% abundance each, have spin $\frac{1}{2}$ and the remaining 76% of the distribution of stable isotopes have zero nuclear magnetic moments. The proton spectrum of this compound in inert media below $-60°$ contains two doublets due to the ^{111}Cd and ^{113}Cd proton coupling, respectively, flanking a single peak due to the methyls on all the other isotopic species containing cadmium with zero nuclear magnetic moment. At higher temperatures, these resonances coalesce due to an intermolecular exchange process

$$CH_3Cd^*CH_3 + CH_3^*CdCH_3^* \rightleftarrows CH_3Cd^*CH_3^* + CH_3CdCH_3^* \qquad (6\text{-}151)$$

proposed to take place via the doubly bridged transition state **2**.

$$
\begin{array}{c}
CH_3 \\
\diagup \quad \diagdown \\
CH_3{-}Cd \qquad Cd{-}CH_3 \\
\diagdown \quad \diagup \\
CH_3
\end{array}
$$

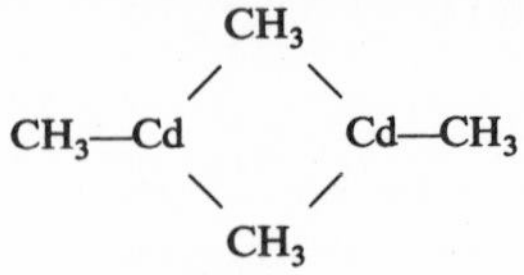

2

The problem has been treated as a five site exchanging system involving uncoupled species using the Kubo–Anderson–Sack formalism and does not properly take account of the mechanism of the exchange process. A more realistic approach to calculating NMR lineshapes for dimethyl-cadmium is to use the density matrix formalism neglecting the off diagonal elements of the spin Hamiltonian. One can further simplify matters by simulating the two methyl groups of dimethylcadmium as a single proton. This will allow us to calculate the proton spectrum though not that for ^{111}Cd or ^{113}Cd.

The species, their simulations, and labels for the lineshape calculation are listed in the following table:

Species	Simulation	Label
^{111}Cd(CH$_3$)$_2$	^{111}CdH	111
^{113}Cd(CH$_3$)$_2$	^{113}CdH	113
Cd(CH$_3$)$_2^a$	^{N}CdH	N

[a] Cadmium isotopes with zero magnetic moment

Since the cadmium and proton transitions do not mix we can calculate the proton absorption separately. NMR parameters needed are

	Abbreviation
$J(^{111}$Cd,H$)$	J_{111}
$J(^{113}$Cd,H$)$	J_{113}

and the proton shift ω_{0H}. As before,

$$\Delta\omega_H = \omega - \omega_{0H} \qquad (6\text{-}152)$$

and one phenomenological linewidth parameter $1/T$ takes care of instrumental and field effects. Product functions of the different simulated species are given in the following table together with their labels.

$^{111}Cd(CH_3)_2$		$^{113}Cd(CH_3)_2$		$Cd(CH_3)_2$	
ϕ_i^{111CdH}	i	ϕ_i^{113CdH}	i	ϕ_i^{H}	i
$\alpha\alpha$	1	$\alpha\alpha$	5	α	9
$\alpha\beta$	2	$\alpha\beta$	6	β	10
$\beta\alpha$	3	$\beta\alpha$	7		
$\beta\beta$	4	$\beta\beta$	8		

Finally, the absorption for the real system is obtained by summing over all density matrix elements, $\Delta m = +1$ (for proton spins) as given in

$$\text{Abs} = -0.76(6c)\,\text{Im}\big[\tilde{\rho}^{N}_{9,\,10}\big] - 0.12(6c)\,\text{Im}\big[\tilde{\rho}^{111}_{1,\,2} + \tilde{\rho}^{111}_{3,\,4} + \tilde{\rho}^{113}_{5,\,6} + \tilde{\rho}^{113}_{7,\,8}\big], \qquad (6\text{-}153)$$

where 6c is the total concentration of hydrogen in dimethyl cadmium.

The exchange processes which must be taken into account are listed together with cadmium mass numbers as

$$CH_3Cd^*CH_3 + CH_3^*CdCH_3^* \underset{}{\overset{k}{\rightleftarrows}} CH_3Cd^*CH_3^* + CH_3^*CdCH_3, \qquad (6\text{-}154)$$

111	111	111	111,	(6-154a)
111	113	111	113,	(6-154b)
113	113	113	113,	(6-154c)
111	N	111	N,	(6-154d)
113	N	113	N,	(6-154e)

Note that the exchange collision of two CH_3CdCH_3 molecules containing cadmiums of zero magnetic moment does not effect the NMR lineshape, and so need not be listed here. We assume that isotope effects among rate constants for Eq. (6-154a)–(6-154e) are too small to detect with the NMR method. The exchange processes (6-154a)–(6-154e) will be simulated as (6-156a)–(6-156e), listed with cadmium isotope masses, as

$$Cd^*H^* + CdH \overset{k'}{\rightleftarrows} Cd^*H + CdH^*, \qquad (6\text{-}155)$$

111	111	111	111,	(6-156a)
111	113	111	113,	(6-156b)
113	113	113	113,	(6-156c)
111	N	111	N,	(6-156d)
113	N	113	N.	(6-156e)

Recalling that

$$1/\tau_{\text{sp, ex}} = R_{\text{ex}}/\,(\text{sp}), \qquad (6\text{-}157)$$

the reciprocal mean lifetimes for the different simulated dimethylcadmium species between successive exchanges (6-119a)–(6-119e) are given by

$$1/\tau_{111,a} = 0.12k'(\text{CdH}), \tag{6-158a}$$

$$1/\tau_{111,b} = 0.12k'(\text{CdH}), \tag{6-158b}$$

$$1/\tau_{113,b} = 0.12k'(\text{CdH}), \tag{6-158c}$$

$$1/\tau_{113,c} = 0.12k'(\text{CdH}), \tag{6-158d}$$

$$1/\tau_{111,d} = 0.76k'(\text{CdH}), \tag{6-158e}$$

$$1/\tau_{113,d} = 0.76k'(\text{CdH}), \tag{6-158f}$$

where we have used the fact that $^{111}CdMe_2$ and $^{113}CdMe_2$ each comprise 12% of the total dimethylcadmium concentration ($CdMe_2$), and of course total ("CdH") is identical to total ($Cd(CH_3)_2$).

Using the procedures of Chapter V in conjunction with the steps in the simulated mechanism (6-156a)–(6-156e), elements of $E\tilde{\rho}$ are evaluated for all proton $\Delta m = 1$ transitions in all species present [see (6-156)] in the exchanging system.

Consider just the simulated reaction (6-156e),

$$\begin{array}{lcccc} & \text{Cd*H*} + & \text{CdH} \overset{k'}{\rightleftarrows} & \text{Cd*H} + & \text{CdH*}, \\ i \text{ of } \phi_i: & 113 & \text{N} & 113 & \text{N} \\ & ab & c & ac & b \end{array} \tag{6-156e}$$

where 113 and N are cadmium isotope labels and i is the label of the product function ϕ_i.

Then, $E\tilde{\rho}^{113}$ just for reaction (6-156e) is

$$\left[E\tilde{\rho}^{113}\right]_{\text{ab, a'b'}} = \frac{1}{\tau_{113,\text{e}}} \sum_{\text{c}} \tilde{\rho}^{113}_{\text{ac, a'c}} \tilde{\rho}^{\text{N}}_{\text{b, b'}} - \frac{1}{\tau_{113,\text{e}}} \tilde{\rho}^{113}_{\text{ab, a'b'}}. \tag{6-159}$$

Next, we evaluate $[E\tilde{\rho}^{\text{N}}]_{9,10}$ by means of the Permutation of Indices (PI) method, noting that this involves only exchange steps (6-119d) and (6-119e) to obtain

$$\begin{aligned} \left[E\tilde{\rho}^{\text{N}}\right]_{9,10} &= (1/\tau_{\text{N,d}})\left[\tilde{\rho}^{111}_{1,2} + \tilde{\rho}^{111}_{3,4} - \tilde{\rho}^{\text{N}}_{9,10}\right] \\ &\quad + (1/\tau_{\text{N,e}})\left[\tilde{\rho}^{113}_{5,6} + \tilde{\rho}^{113}_{7,8} - \tilde{\rho}^{\text{N}}_{9,10}\right] \\ &= 0.12k'(\text{CdH})\left[\tilde{\rho}^{111}_{1,2} + \tilde{\rho}^{111}_{3,4} + \tilde{\rho}^{113}_{5,6} + \tilde{\rho}^{113}_{7,8} - 2\tilde{\rho}^{\text{N}}_{9,10}\right]. \end{aligned} \tag{6-160}$$

In this way, we derive the proton portion of the density matrix equations for the exchanging system described by Eqs. (6-156a)–(6-156e). Recall that since cadmium and proton transitions do not mix, the proton absorption can be solved for separately.

To make the connection between the simulated exchanging system (6-155) and the proposed mechanism it is necessary to replace

$$1/\tau_{\mathrm{CdH}} \quad \text{by} \quad 2/\tau_{\mathrm{Cd(CH_3)_2}} \tag{6-161}$$

or

$$k'(\mathrm{CdH}) \quad \text{by} \quad 2k(\mathrm{Cd(CH_3)_2}). \tag{6-162}$$

The result of the treatment just described, including the conversion expressed by (6-162), is the set of coupled density matrix equations for the $\Delta m = 1$ proton transitions of system (6-155) given as

$$\begin{bmatrix} \begin{pmatrix} i\,\Delta\omega_{\mathrm{H}} - (i\,J_{111}/2) \\ - (1/T) - 3.76k(\mathrm{c}) \end{pmatrix} & 0.24k(\mathrm{c}) & 0.24k(\mathrm{c}) & 0.24k(\mathrm{c}) & 1.52k(\mathrm{c}) \\ 0.24k(\mathrm{c}) & \begin{pmatrix} i\,\Delta\omega_{\mathrm{H}} - (i\,J_{113}/2) \\ - (1/T) - 3.76k(\mathrm{c}) \end{pmatrix} & 0.24k(\mathrm{c}) & 0.24k(\mathrm{c}) & 1.52k(\mathrm{c}) \\ 0.24k(\mathrm{c}) & 0.24k(\mathrm{c}) & \begin{pmatrix} i\,\Delta\omega_{\mathrm{H}} + (i\,J_{111}/2) \\ - (1/T) - 3.76k(\mathrm{c}) \end{pmatrix} & 0.24k(\mathrm{c}) & 1.52k(\mathrm{c}) \\ 0.24k(\mathrm{c}) & 0.24k(\mathrm{c}) & 0.24k(\mathrm{c}) & \begin{pmatrix} i\,\Delta\omega_{\mathrm{H}} + (i\,J_{113}/2) \\ - (1/T) - 3.76k(\mathrm{c}) \end{pmatrix} & 1.52k(\mathrm{c}) \\ 0.48k(\mathrm{c}) & 0.48k(\mathrm{c}) & 0.48k(\mathrm{c}) & 0.48k(\mathrm{c}) & \begin{pmatrix} i\,\Delta\omega_{\mathrm{H}} - (1/T) \\ -0.96k(\mathrm{c}) \end{pmatrix} \end{bmatrix} \begin{bmatrix} \tilde{\rho}^{111}_{1,2} \\ \tilde{\rho}^{113}_{5,6} \\ \tilde{\rho}^{111}_{3,4} \\ \tilde{\rho}^{113}_{7,8} \\ \tilde{\rho}^{\mathrm{N}}_{9,10} \end{bmatrix}$$

$$= \frac{i\beta'}{4} \begin{bmatrix} 1 \\ 1 \\ 1 \\ 1 \\ 2 \end{bmatrix} \tag{6-163}$$

Matrix (6-160) converts to the Kubo–Anderson–Sack form on replacing the ρ_{col} by

$$\tfrac{1}{2}\tilde{\rho}^{111}_{1,2s}, \quad \tfrac{1}{2}\tilde{\rho}^{113}_{5,6s}, \quad \tfrac{1}{2}\tilde{\rho}^{111}_{3,4s}, \quad \tfrac{1}{2}\tilde{\rho}^{111}_{7,8s}, \quad \text{and} \quad \tilde{\rho}^{\mathrm{N}}_{9,10s} \tag{6-164}$$

listed in order from top to bottom.

One might argue that a more realistic simulation of the exchange process would be to represent $\mathrm{Cd(CH_3)_2}$ as a three spin molecule, Cd"$\mathrm{H_2}$", one for the cadmium and one each for the methyls. Then, all possible mutual exchanges of two simulated methyls among all pairs of cadmium isotopic species would have to be considered. Note that we have to take account of four ways of exchanging two methyls between two dimethylcadmiums, see the connecting lines in Fig. 6-5. This complicated calculation of the proton lineshape for the exchanging Cd"$\mathrm{H_2}$" system (compared to Cd"H") turns out to be a fruitless waste of time because the result after equivalence

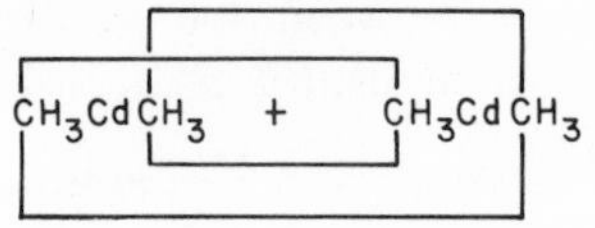

Fig. 6-5. Four ways two methyl groups can mutually exchange between two dimethylcadmium molecules.

factoring is exactly the same as for "CdH", with the exception that k' in (6-158) is replaced by $2k$. The reader who does not believe this should do the problem as an exercise!

The example of the dimethylcadmium exchanging system has been presented to show how a fairly complicated system can be handled and what sorts of approximations are appropriate.

f. Strongly Coupled Exchanging Systems

(1) First Order Rearrangements[17–22]

When shifts and coupling constants are of the same order, the appearance of the spectrum, compared to first order systems, is much more sensitive to small changes in the NMR parameters, minor variations in the exchange mechanism and such effects as relative signs of coupling constants. It is an important principle that kinetic analysis of equilibrium exchanging systems using the NMR lineshape method is in every way a more useful technique when tightly coupled systems are studied; the results contain a wealth of mechanistic detail, together with information on relaxation and other NMR parameters which are often not available from first order systems.

For tightly coupled systems we require the full nuclear spin Hamiltonian given as

$$\mathcal{H} = \sum_s (\omega_{0s} - \omega) I_s^z + \sum_{s>t} J_{s,t}(I_s^x I_t^x + I_s^y I_t^y + I_s^z I_t^z) + \omega_1 \sum_s I_s^x \,. \tag{6-165}$$

The remaining procedure for deriving the density matrix equations follows the methods of Chapter V for the exchange terms $E\tilde{\rho}$ and the low power approximations discussed above in this chapter.

The principle use to which these lineshapes of strongly coupled systems have been put is to measure the rates of kinetically first order processes such as rotation, pseudo rotation, Berry rotation, inversion, and other intramolecular rearrangements. A large fraction of the activation parameters for conformational interconversion comes from NMR lineshape data. The literature is filled with examples of this kind of work including a recently published book.[23] In all these studies of unimolecular processes the exchange contribution to the density matrix equations comes in two forms. Degenerate rearrangements,

$$A \rightleftarrows A, \tag{6-166}$$

involve no change in the Hamiltonian but the arrangement of spins in A becomes scrambled as a result of the reorganization process. Then, $E\rho$ is given by

$$\left[E\tilde{\rho}^{A}\right]_{a,a'} = (1/\tau_A)\left[\tilde{\rho}^{A}_{b,b'} - \tilde{\rho}^{A}_{a,a'}\right], \tag{6-167}$$

where the order of the spins in the product functions ϕ_a and $\phi_{a'}$ labels their positions in the molecular species A. The same position label order is preserved in ϕ_b and $\phi_{b'}$. In unimolecular interconversion

$$\begin{matrix} A \rightleftarrows B \\ \phi_a \quad \phi_a \end{matrix} \tag{6-168}$$

the wave function remains the same, but the Hamiltonian switches from $\mathcal{H}^A$ to $\mathcal{H}^B$ so an element of $E\tilde{\rho}^A$ is

$$\left[E\tilde{\rho}^{A}\right]_{a,a'} = (1/\tau_A)\left[\tilde{\rho}^{B}_{a,a'} - \tilde{\rho}^{A}_{a,a'}\right]. \tag{6-169}$$

(2) THE AB EXCHANGING SYSTEM[24–28]

There are numerous examples where intramolecular rearrangements, such as inversion or rotation processes, bring about the averaging of an AB quartet into a single line as described by

$$\begin{matrix} & AH^1BH^2 & \overset{1/\tau_1}{\rightleftarrows} & AH^2BH^1, \\ i \text{ of } \phi_i^{AB}: & ab & & ba \end{matrix} \tag{6-170}$$

where A and B label environments, the numbers label the single hydrogens, and the lower case letters label the spin functions.

A second exchange process involves a mutual switching of two non-equivalent protons between two identical AB molecules

$$\begin{matrix} & AH^1BH^2 + AH^3BH^4 & \overset{1/\tau_2}{\rightleftarrows} & AH^4BH^2 + AH^3BH^1 \\ i \text{ of } \phi_i^{AB}: & ab \qquad cd & & db \qquad ca \end{matrix} \tag{6-171a}$$

$$\begin{matrix} & AH^1BH^2 + AH^3BH^4 & \overset{1/\tau_2}{\rightleftarrows} & AH^1BH^3 + AH^2BH^4. \\ i \text{ of } \phi_i^{AB}: & ab \qquad cd & & ac \qquad bd \end{matrix} \tag{6-171b}$$

Both the intramolecular exchange (6-170) and the intermolecular exchanges (6-171a) and (6-171b) have the effect of averaging the AB quartet to a single line. For completeness we also mention the mutual exchange of two equivalent protons between two AB molecules

$$\begin{matrix} & AH^1BH^2 + AH^3BH^4 & \overset{1/\tau_2'}{\rightleftarrows} & AH^1BH^4 + AH^3BH^2. \\ i \text{ of } \phi_i^{AB}: & ab \qquad cd & & ad \qquad cb \end{matrix} \tag{6-172}$$

Here, exchange has the effect of averaging the AB coupling constant but not the shift, thus averaging a four line AB multiplet into a doublet. An example of such an exchange process (6-172) could be the exchange of hydroxyl protons between two benzhydrol molecules

$$\phi_2\mathrm{CH\overset{*}{O}H^*} + \phi_2\mathrm{CHOH} \rightleftarrows \phi_2\mathrm{CH\overset{*}{O}H} + \phi_2\mathrm{CHOH^*}. \tag{6-173}$$

Let us first compare the NMR lineshape behavior of an AB system undergoing the first two kinds of exchange behavior, intramolecular (6-170) and intermolecular exchange of nonequivalent protons (6-171). Although both processes bring about averaging of the AB quartet, we shall show how the details of the lineshape changes are different.

The density matrix equations required to obtain the NMR lineshapes for the three exchanging systems described above have terms *common* to all of the systems and differ only in the $E\tilde{\rho}^{AB}$ exchange-dependent contribution from one process to another. The exchange-independent terms come from $i[\tilde{\rho}^{AB}, \overline{\mathcal{H}}] - \rho/T$.

The Hamiltonian for an AB system in the rotating frame for this system is

$$\overline{\mathcal{H}} = (\omega_{0A} - \omega)I_A^z + (\omega_{0B} - \omega)I_B^z + J_{AB}I_A \cdot I_B + \omega_1(I_A^x + I_B^x), \tag{6-174}$$

we label the product functions ϕ_i in the order

ϕ_i^{AB}	i
$\alpha\alpha$	1
$\alpha\beta$	2
$\beta\alpha$	3
$\beta\beta$	4,

and we shall use phenomenological linewidth parameters $1/T_A$ and $1/T_B$. Also, to avoid confusion, we label the preexchange lifetimes of AB as shown above τ_1 for (6-170), τ_2 for (6-171), and $\tau_{2'}$ for (6-172). They are related to the rate constants as given by

$$1/\tau_1 = k_1, \tag{6-175}$$

$$1/\tau_2 = k_2(\mathrm{AB}), \tag{6-176}$$

$$1/\tau_{2'} = k_{2'}(\mathrm{AB}). \tag{6-177}$$

The exchange contribution $E\tilde{\rho}^{AB}$ to the density matrix equations for intramolecular exchange (6-170) is

$$\left[E\tilde{\rho}^{AB}\right]_{ab,\,a'b'} = (1/\tau_1)\left[\tilde{\rho}^{AB}_{ba,\,b'a'} - \tilde{\rho}^{AB}_{ab,\,a'b'}\right] \tag{6-178}$$

according to the prescriptions of Chapter V.

Using the PI method for intermolecular exchange (6-171) we obtain $E\tilde{\rho}^{AB}$ as

$$\left[E\tilde{\rho}^{AB}\right]_{ab,\,a'b'} = (1/\tau_2)\sum\left[\tilde{\rho}^{AB}_{db,\,db'}\tilde{\rho}^{AB}_{ca,\,ca'} + \tilde{\rho}^{AB}_{ac,\,a'c}\tilde{\rho}^{AB}_{bd,\,b'd}\right] - (1/\tau_2)2\tilde{\rho}^{AB}_{ab,\,a'b'}. \tag{6-179}$$

Just to show how Eq. (6-179) works out here is the $\langle 1|E\tilde{\rho}^{\mathrm{AB}}|2\rangle$ element

$$\left[E\tilde{\rho}^{\mathrm{AB}}\right]_{1,2} = (1/\tau_2)\left\{\left[\tilde{\rho}^{\mathrm{AB}}_{1,1} + \tilde{\rho}^{\mathrm{AB}}_{3,3}\right]\left[\tilde{\rho}^{\mathrm{AB}}_{1,2} + \tilde{\rho}^{\mathrm{AB}}_{3,4}\right] + \left[\tilde{\rho}^{\mathrm{AB}}_{1,1} + \tilde{\rho}^{\mathrm{AB}}_{2,2}\right]\left[\tilde{\rho}^{\mathrm{AB}}_{1,3} + \tilde{\rho}^{\mathrm{AB}}_{2,4}\right] - 2\tilde{\rho}^{\mathrm{AB}}_{1,2}\right\}. \quad (6\text{-}180)$$

Linearization, as discussed in Chapter V, reduces (6-180) to

$$\left[E\tilde{\rho}^{\mathrm{AB}}\right]_{1,2} = (1/2\tau_2)\left[\tilde{\rho}^{\mathrm{AB}}_{3,4} + \tilde{\rho}^{\mathrm{AB}}_{1,3} + \tilde{\rho}^{\mathrm{AB}}_{2,4}\right] - (3/2\tau_2)\tilde{\rho}^{\mathrm{AB}}_{1,2}, \quad (6\text{-}181)$$

where each diagonal element of $\tilde{\rho}^{\mathrm{AB}}$ has been replaced by $\frac{1}{4}$.

The density matrix equations which describe (6-170) and (6-171) are of the form

$$i\left[\tilde{\rho}^{\mathrm{AB}}, \mathcal{H}^{\mathrm{AB}}\right] + R\tilde{\rho}^{\mathrm{AB}} + E\tilde{\rho}^{\mathrm{AB}} = 0. \quad (6\text{-}182)$$

As a further example, here is the $\langle 1||2\rangle$ element of (6-182) for intramolecular exchange (6-170)

$$\tilde{\rho}^{\mathrm{AB}}_{1,2}\left[i\left(\Delta\omega_{0\mathrm{B}} - \frac{J_{\mathrm{AB}}}{2}\right) - \frac{1}{T_{\mathrm{B}}} - \frac{1}{\tau_1}\right] + \tilde{\rho}^{\mathrm{AB}}_{1,3}\left[\frac{iJ_{\mathrm{AB}}}{2} + \frac{1}{\tau_1}\right] = \frac{i\beta'}{4}, \quad (6\text{-}183)$$

where $1/T_{\mathrm{B}}$ is the instrumental linewidth parameter. The coupled equations in matrix form for the two exchanging systems are written as

$$\begin{bmatrix} \\ i\omega\mathcal{I} + A \\ \\ \end{bmatrix}\begin{bmatrix} \tilde{\rho}^{\mathrm{AB}}_{1,2} \\ \tilde{\rho}^{\mathrm{AB}}_{1,3} \\ \tilde{\rho}^{\mathrm{AB}}_{2,4} \\ \tilde{\rho}^{\mathrm{AB}}_{3,4} \end{bmatrix} = \frac{i\beta'}{4}\begin{bmatrix} 1 \\ 1 \\ 1 \\ 1 \end{bmatrix}, \quad (6\text{-}184)$$

followed by the coefficient matrix A for intramolecular exchange (6-170) and for intermolecular exchange (6-171), shown, respectively, as

$$\begin{bmatrix} \begin{pmatrix} i(-\omega_{0\mathrm{B}} - J/2) \\ -(1/T_{\mathrm{B}}) - 1/\tau_1 \end{pmatrix} & (iJ/2) + 1/\tau_1 & 0 & 0 \\ (iJ/2) + 1/\tau_1 & \begin{pmatrix} i(-\omega_{0\mathrm{A}} - J/2) \\ -(1/T_{\mathrm{A}}) - 1/\tau_1 \end{pmatrix} & 0 & 0 \\ 0 & 0 & \begin{pmatrix} i(-\omega_{0\mathrm{A}} + J/2) \\ -(1/T_{\mathrm{A}}) - 1/\tau_1 \end{pmatrix} & -(iJ/2) + 1/\tau_1 \\ 0 & 0 & -(iJ/2) + 1/\tau_1 & \begin{pmatrix} i(-\omega_{0\mathrm{B}} + J/2) \\ -(1/T_{\mathrm{B}}) - 1/\tau_1 \end{pmatrix} \end{bmatrix}$$

(6-185)

$$\begin{bmatrix} \begin{pmatrix} i(-\omega_{0B} - J_{AB}/2) \\ -(1/T_B) - 3/2\tau_2 \end{pmatrix} & (i\,J_{AB}/2) + 1/2\tau_2 & 1/2\tau_2 & 1/2\tau_2 \\ (i\,J/2) + 1/2\tau_2 & \begin{pmatrix} i(-\omega_{0A} - J_{AB}/2) \\ -(1/T_A) - 3/2\tau_2 \end{pmatrix} & 1/2\tau_2 & 1/2\tau_2 \\ 1/2\tau_2 & 1/2\tau_2 & \begin{pmatrix} i(-\omega_{0A} + J_{AB}/2) \\ -(1/T_A) - 3/2\tau_2 \end{pmatrix} & -(i\,J_{AB}/2) + 1/2\tau_2 \\ 1/2\tau_2 & 1/2\tau_2 & -(i\,J_{AB}/2) + 1/2\tau_2 & \begin{pmatrix} i(-\omega_{0B} + J_{AB}/2) \\ -(1/T_B) - 3/2\tau_2 \end{pmatrix} \end{bmatrix} \tag{6-186}$$

It is instructive to examine matrices (6-185) and (6-186). Note first how the arrangement of off diagonal elements in $1/\tau$ is entirely different in the two matrices. It would be understandable that the lineshapes might be distinguishable. It is interesting to note how the density matrix equations for intramolecular exchange can be separated into two sets of two each, which should reduce the amount of computer time used for numerical calculations. Finally, we now display the coefficient matrix for intermolecular exchange for equivalent protons in (6-172). Following the PI method of Chapter V, the coefficient matrix A comes out as

$$\begin{bmatrix} \begin{pmatrix} i(\omega_{0B} - J_{AB}/2) \\ -(1/T) - 1/2\tau_{2'} \end{pmatrix} & i\,J_{AB}/2 & 0 & 1/2\tau_{2'} \\ iJ_{AB}/2 & \begin{pmatrix} i(\omega_{0A} - J_{AB}/2) \\ -(1/T) - 1/2\tau_{2'} \end{pmatrix} & 1/2\tau_{2'} & 0 \\ 0 & 1/2\tau_{2'} & \begin{pmatrix} i(\omega_{0A} + J_{AB}/2) \\ -(1/T) - 1/2\tau_{2'} \end{pmatrix} & -i\,J_{AB}/2 \\ 1/2\tau_{2'} & 0 & -iJ_{AB}/2 & \begin{pmatrix} i(\omega_{0B} + J_{AB}/2) \\ -(1/T) - 1/2\tau_{2'} \end{pmatrix} \end{bmatrix} \tag{6-187}$$

Typical lineshapes for these AB systems undergoing exchange processes (6-170) and (6-171) are displayed in Fig. 6-6. Note how over a wide range of τ_{AB} values, the two sets of curves are significantly different and could not be confused.

(3) The ABC and ABX Exchanging Systems

An example of a system whose lineshapes are very sensitive to exchange processes is the three spin ABC system. We consider this as a tightly

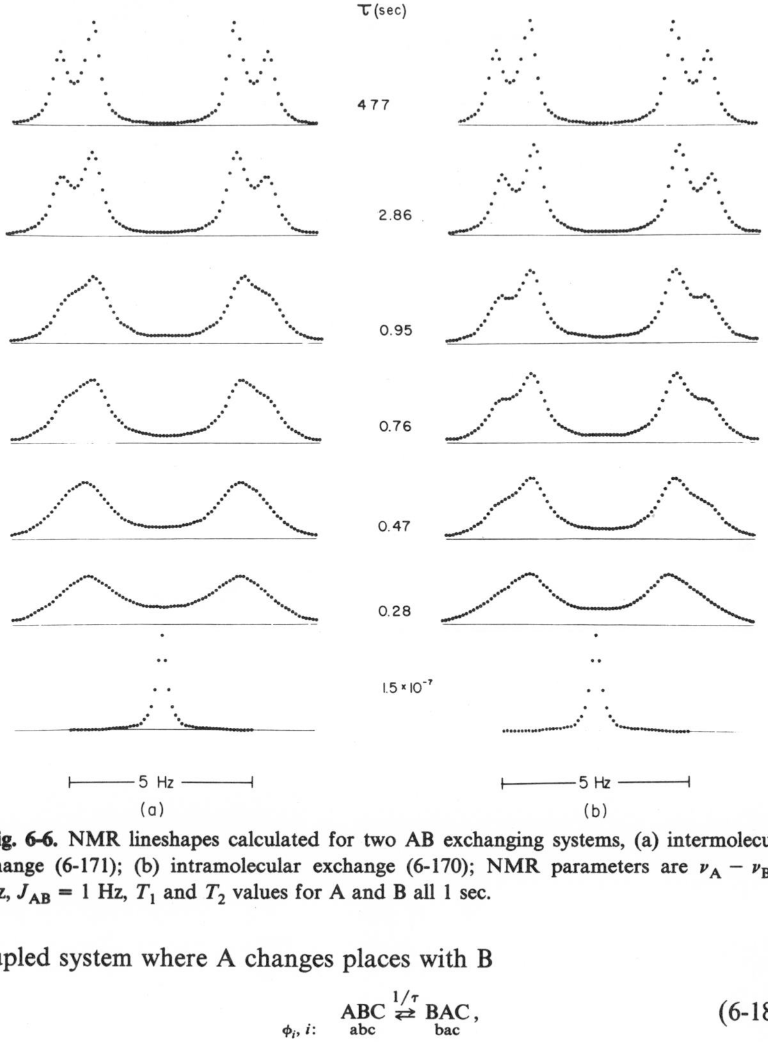

Fig. 6-6. NMR lineshapes calculated for two AB exchanging systems, (a) intermolecular exchange (6-171); (b) intramolecular exchange (6-170); NMR parameters are $\nu_A - \nu_B = 5$ Hz, $J_{AB} = 1$ Hz, T_1 and T_2 values for A and B all 1 sec.

coupled system where A changes places with B

$$\underset{\phi_i,\ i:\quad \text{abc}}{\text{ABC}} \overset{1/\tau}{\rightleftarrows} \underset{\text{bac}}{\text{BAC}}, \tag{6-188}$$

the order indicates the position and ABC is chemically identical to BAC. Below we shall also discuss the ABX system where

$$\begin{aligned} \omega_{0A} - \omega_{0B} &\ll \omega_{0A} - \omega_{0X}\,, \omega_{0B} - \omega_{0X}\,, \\ J_{AX}\,, J_{BX} &\ll \omega_{0A} - \omega_{0X}\,, \omega_{0B} - \omega_{0X}\,, \end{aligned} \tag{6-189}$$

and considerable simplification of the lineshape calculation is possible. In this connection there will be some consideration of relative signs of coupling constants.

The following parameters describe the ABC exchanging system. The product functions ϕ_i are listed as

ϕ_i	i	ϕ_i	i
$\alpha\alpha\alpha$	1	$\beta\beta\alpha$	5
$\alpha\alpha\beta$	2	$\beta\alpha\beta$	6
$\alpha\beta\alpha$	3	$\alpha\beta\beta$	7
$\beta\alpha\alpha$	4	$\beta\beta\beta$	8

with chemical shifts

$$\omega_{0A}, \omega_{0B}, \text{ and } \omega_{0C}$$

relative to the observe frequency ω

$$\Delta\omega_A = \omega - \omega_{0A}, \qquad \Delta\omega_B = \omega - \omega_{0B}, \qquad \Delta\omega_C = \omega - \omega_{0C}, \quad (6\text{-}190)$$

coupling constants

$$J_{AB}, J_{AC}, J_{BC},$$

and inhomogeneity linewidth parameters $1/T_A$, $1/T_B$, $1/T_C$.

The Hamiltonian is

$$\begin{aligned}\overline{\mathcal{H}} = {} & (\omega_{0A} - \omega)I_A^z + (\omega_{0B} - \omega)I_B^z + (\omega_{0C} - \omega)I_C^z \\ & + J_{AB}I_A \cdot I_B + J_{AC}I_A \cdot I_C + J_{BC}I_B \cdot I_C \\ & + \not{\omega}_1\left[I_A^x + I_B^x + I_C^x\right],\end{aligned} \quad (6\text{-}191)$$

and we shall need to solve for the density matrix elements $\tilde{\rho}_{i,k}$ listed here as follows:

i, k	transition	i, k	transition
2, 6	A	1, 2	C
7, 8	A	4, 6	C
2, 7	B	3, 7	C
6, 8	B	5, 8	C
1, 4	A	3, 6	combination
3, 5	A	2, 5	combination
1, 3	B	4, 7	combination
4, 5	B		

This time, since the exchange process only happens inside the ABC molecule, we can leave off the species label. Exchange of A with B gives a $E\tilde{\rho}$ term which is

$$\left[E\tilde{\rho}\right]_{abc,\, a'b'c'} = (1/\tau)\left[\tilde{\rho}_{bac,\, b'a'c'} - \tilde{\rho}_{abc,\, a'b'c'}\right]. \quad (6\text{-}192)$$

Taking the 15 $\langle i|\,|k\rangle$ elements, listed above, of the density matrix equation gives the matrix equation in the form

$$(\mathcal{I}i(\omega - \omega_0) + A)\tilde{\rho} = (i\beta'/8)B, \quad (6\text{-}193a)$$

where the term $A\tilde{\rho}$ in (6-193a) is displayed in (6-193b) on pp. 108 and 109.† The absorption comes from the imaginary part of the sum of

$$\mathrm{Abs}(\omega) = -\mathrm{Im}\sum_{i,k} \tilde{\rho}_{i,k}^{\mathrm{ABC}}, \tag{6-194}$$

where

$$|I^{+}|k\rangle = +|i\rangle, \qquad m_i - m_k = +1. \tag{6-195}$$

The sum is over the transitions listed in (6-191) but does *not include* the combination matrix elements.

The ABC example brings up an instructive comment about combination transitions. Their matrix elements, here $\tilde{\rho}_{3,6}$, $\tilde{\rho}_{2,5}$, and $\tilde{\rho}_{4,5}$, do not contribute directly to the absorption, and the corresponding elements of $\Sigma_s I_s^y$ in the B column of the density matrix equation are zero. One might conclude from this that the calculated absorption does not include the combination transitions. Actually, their absorption comes indirectly via coupling to other transitions which do contribute directly to the absorption.

In the event that the C proton is shifted far from A and B, which are close, then ABC becomes ABX, and the calculation of the lineshape becomes considerably simplified. Under these circumstances, J_{AX} and J_{BX} can be dropped wherever they occur in off-diagonal elements. The matrix can then be separated into two 4×4 matrices for the AB resonance (boxed), each of which describes a pseudo-AB quartet, see below, and a 7×7 matrix which includes the X and combination transitions. Therefore, the AB lineshape can be calculated separately from the rest of the spectrum.

If the combination transitions are sufficiently weak compared to the rest of the spectrum their density matrix elements can be dropped from the coupled equations leaving a 4×4 matrix for the X resonance.

A simplified view of the AB part of an ABX spectrum is that each spin state of X generates a new pseudo-AB quartet from the original AB system.[29] An example of the AB part of an ABX system is shown in Fig. 6-7 where the two quartets are drawn with thick and thin lines, respectively. We can assume that during an exchange of A with B the spin state

†The terms in (6-193a) and (6-193b) are abbreviated as follows:

$$\mathrm{A} = i(\omega_0 - \omega_{0\mathrm{A}}), \qquad \mathrm{B} = i(\omega_0 - \omega_{0\mathrm{B}}), \qquad \mathrm{C} = i(\omega_0 - \omega_{0\mathrm{C}}),$$
$$\mathrm{AB} = (i/2)J_{\mathrm{AB}}, \qquad \mathrm{AC} = (i/2)J_{\mathrm{AC}}, \qquad \mathrm{BC} = (i/2)J_{\mathrm{BC}}$$

and the $\tilde{\rho}_{\mathrm{col}}$ shows the order of equations.

$$
\left[\begin{array}{cccc|cccc}
A + AC - BC - T^{-1} - \tau^{-1} & 0 & AB + \tau^{-1} & 0 & 0 & 0 & 0 & 0 \\
0 & A + AB + AC - T^{-1} - \tau^{-1} & 0 & -AB + \tau^{-1} & 0 & 0 & 0 & 0 \\
AB + \tau^{-1} & 0 & B - AB + BC - T^{-1} - \tau^{-1} & 0 & 0 & 0 & 0 & 0 \\
0 & -AB + \tau^{-1} & 0 & A - AB - AC - T^{-1} - \tau^{-1} & 0 & 0 & 0 & 0 \\
\hline
0 & 0 & 0 & 0 & A - AB - AC - T^{-1} - \tau^{-1} & 0 & AB + \tau^{-1} & 0 \\
0 & 0 & 0 & 0 & 0 & A + AB - AC - T^{-1} - \tau^{-1} & 0 & -AB + \tau^{-1} \\
0 & 0 & 0 & 0 & AB + \tau^{-1} & 0 & B - AB - BC - T^{-1} - \tau^{-1} & 0 \\
0 & 0 & 0 & 0 & 0 & -AB + \tau^{-1} & 0 & B - AB - BC - T^{-1} - \tau^{-1} \\
\hline
0 & 0 & 0 & 0 & AC & 0 & BC & 0 \\
-AC & 0 & 0 & 0 & 0 & 0 & 0 & BC \\
0 & 0 & -BC & 0 & 0 & AC & 0 & 0 \\
0 & -AC & 0 & -BC & 0 & 0 & 0 & 0 \\
-BC & -BC & 0 & 0 & 0 & BC & 0 & 0 \\
BC & 0 & AC & 0 & 0 & -BC & 0 & -AC \\
0 & 0 & -AC & 0 & 0 & 0 & 0 & AC
\end{array}\right.
$$

---

of proton X does not change. Thus, the effect of an intramolecular AB exchange is to average the A resonance with the B resonance in each pseudoquartet separately. Fast exchange collapses the two quartets to lines at their centers separated by

$$\tfrac{1}{2}(J_{AX} + J_{BX}). \tag{6-196}$$

The lineshapes for the two pseudo-AB quartets for different rates of

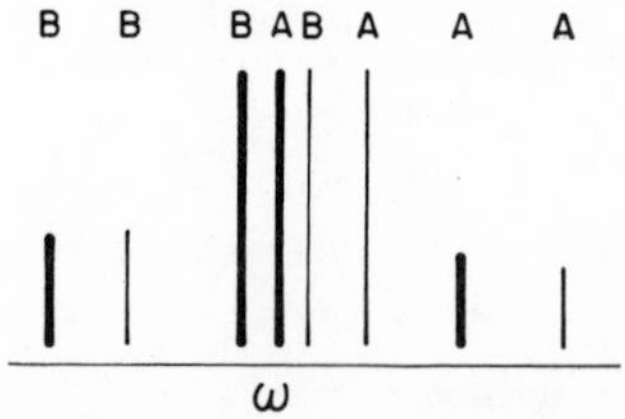

Fig. 6-7. AB part of ABX system showing two pseudo AB quartets using thick and thin lines, respectively.

$$
\begin{array}{ccccccc}
0 & -AC & 0 & 0 & -BC & BC & 0 \\
0 & 0 & 0 & -AC & -BC & 0 & 0 \\
0 & 0 & -BC & 0 & 0 & AC & -AC \\
0 & 0 & 0 & -BC & 0 & 0 & 0 \\
AC & 0 & 0 & 0 & 0 & 0 & 0 \\
0 & 0 & AC & 0 & BC & -BC & 0 \\
BC & 0 & 0 & 0 & 0 & 0 & 0 \\
0 & BC & 0 & 0 & 0 & -AC & AC \\
\hline
C - BC - AC - T^{-1} & 0 & 0 & 0 & 0 & 0 & 0 \\
0 & C - BC + AC - T^{-1} - \tau^{-1} & \tau^{-1} & 0 & -AB & 0 & AB \\
0 & \tau^{-1} & C + BC - AC - T^{-1} - \tau^{-1} & 0 & AB & 0 & -AB \\
0 & 0 & 0 & C + BC + AC - T^{-1} & 0 & 0 & 0 \\
0 & -AB & AB & 0 & A - B + C - T^{-1} - \tau^{-1} & 0 & \tau^{-1} \\
0 & 0 & 0 & 0 & 0 & A + B - C - T^{-1} & 0 \\
0 & AB & -AB & 0 & \tau^{-1} & 0 & -A + B + C - T^{-1} - \tau^{-1}
\end{array}
\Bigg]
\begin{bmatrix}
\bar\rho_{2,6} \\ \bar\rho_{7,8} \\ \bar\rho_{2,7} \\ \bar\rho_{6,8} \\ \bar\rho_{1,4} \\ \bar\rho_{3,5} \\ \bar\rho_{1,3} \\ \bar\rho_{4,5} \\ \bar\rho_{1,2} \\ \bar\rho_{4,6} \\ \bar\rho_{3,7} \\ \bar\rho_{5,8} \\ \bar\rho_{3,6} \\ \bar\rho_{2,5} \\ \bar\rho_{4,7}
\end{bmatrix}
= A\bar\rho
\tag{6-193b}
$$

exchange can be calculated as separate AB systems and then summed to give the AB part of the ABX system.

The procedure for picking out the separate pseudo-AB quartets from the ABX spectrum (uneffected by exchange) is well described in the literature.[30] The separations between the intense lines of each quartet are given by

$$E_{\pm} = \left\{\left[\omega_{0A} - \omega_{0B} \pm \tfrac{1}{2}(J_{AX} - J_{BX})\right]^2 + J_{AB}^2\right\}^{1/2} \tag{6-197}$$

so that when we compare the pseudo-AB system to a real one we see the pseudoshifts to be

$$\omega_{0A} - \omega_{0B} \pm \tfrac{1}{2}(J_{AX} - J_{BX}). \tag{6-198}$$

One exchange process, which was first analyzed by the procedure just given, involved inversion at carbon bonded to magnesium in 2-methylbutylmagnesium halides and related compounds, **3**.[29, 31]

XAB
CH_3CH_2 $CHCH_2MgBr = RMgBr.$
|
CH_3
3

At $-30°$ C, the CH_2Mg proton resonance of **3** consists of the familiar AB part of an ABX spectrum, due to the chirality at C_2 (the penultimate carbon). The AB portion is well resolved from the rest of the spectrum and can therefore be calculated independently. At higher temperatures, increasing rates of inversion at carbon-1 causes the AB shift and coupling constant to progressively average out so that by 90° all that remains for the AB part is a doublet of separation $\frac{1}{2}(J_{AX} + J_{BX})$. This provides an experimental value for the average of these two coupling constants and of their relative signs.

NMR lineshapes for this system were calculated for different rates of AB exchange (caused by inversion) using the full 15 × 15 ABC matrix as well as the pseudo AB quartet method just described. Both methods gave the same results within the accuracy of the experiment.

Later on, the kinetics of inversion were found to be second order in contained Grignard reagent since a plot of $\log(1/\tau)$ versus log (RMgBr) gave a line of slope +1. Since the Grignard reagent was found to be a monomer in the ether used as solvent in the studies, it was concluded that the transition state was dimeric in RMgBr. Thus, on the basis of NMR lineshape data, a mechanism for inversion in organomagnesium coupounds was developed. It was concluded that the transition state could be partially described as **4**.

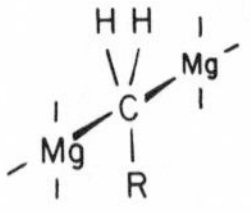

4

Simplifications like the one just described always apply when one multiplet, due to a set of strongly coupled nuclei, is far from others (e.g., AB from X) and where the transitions for the two different multiplets do not mix due to exchange processes. Thus, for the rotating 4-acetyl-1,4-dihydropyridine,

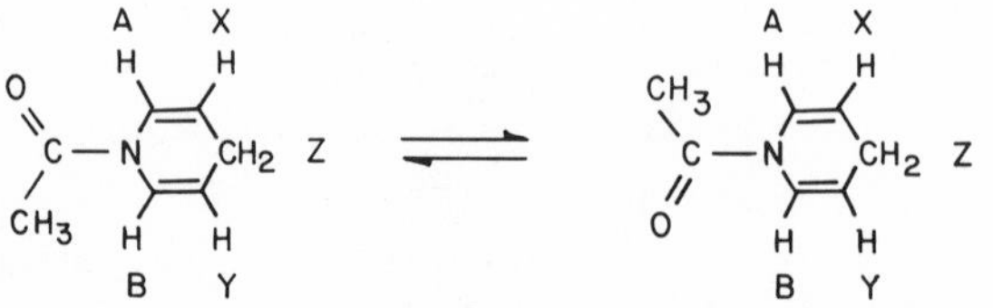

(6-199)

5

fast rotation about the nitrogen carbonyl carbon bond averages the proton resonances of A with B, and X with Y.[32] The ring proton resonance comes in three multiplets that can be calculated separately—AB, XY, and Z_2. Then, in evaluating $i[\rho, \mathcal{H}]_{i,k}$, say for the AB part of the spectrum, one neglects J(XX), (XY), (AZ), (BX), (BY), and (BZ), the weak couplings, wherever they occur in off diagonal elements of the coefficient matrix. This amounts to taking elements only in $I_s^z I_t^z$. No such approximation can be made for J(AB) or J(XY) for these represent strongly coupled pairs of protons.

(4) THE SYSTEM AB + B′ ⇄ AB′ + B

One final example of this section concerns the bimolecular exchange

$$\begin{array}{rcccc} & \mathrm{AB} + & \mathrm{B'} \rightleftarrows & \mathrm{AB'} + & \mathrm{B} \\ i \text{ of } \phi_i = & \mathrm{ab} & \mathrm{c} & \mathrm{ac} & \mathrm{b}, \end{array} \tag{6-200}$$

where A, B, and B′ represent protons in three different environments, two in the AB molecule and one in the outside world. An example of such a process would be the exchange of hydroxyl protons between benzhydrol and water

$$\phi_2\mathrm{CHOH^*} + \mathrm{H_2O} \rightleftarrows \phi_2\mathrm{CHOH} + \mathrm{HOH^*}. \tag{6-201}$$

The Hamiltonians for the two molecules are given as

$$\bar{\mathcal{H}}^{\mathrm{AB}} = (\omega_{0\mathrm{A}} - \omega)I_{\mathrm{A}}^z + (\omega_{0\mathrm{B}'} - \omega)I_{\mathrm{B}}^z + J_{\mathrm{AB}}I_{\mathrm{A}} \cdot I_{\mathrm{B}} + \not{\omega}_1(I_{\mathrm{A}}^x + I_{\mathrm{B}}^x) \tag{6-202}$$

and

$$\bar{\mathcal{H}}^{\mathrm{B}'} = (\omega_{0\mathrm{B}'} - \omega)I_{\mathrm{B}'}^z + \not{\omega}_1 I_{\mathrm{B}'}^x. \tag{6-203}$$

The exchange lifetimes are related to the rate law as given by

$$1/\tau_{\mathrm{AB}} = k(\mathrm{B}'), \qquad 1/\tau_{\mathrm{B}'} = k(\mathrm{AB}) \tag{6-204}$$

and their contributions to the density matrix equations appear as

$$E\tilde{\rho}^{\mathrm{AB}} = (1/\tau_{\mathrm{AB}})\left[\tilde{\rho}^{\mathrm{AB}}(\mathrm{col}) - \tilde{\rho}^{\mathrm{AB}}\right] \tag{6-205}$$

for AB and as

$$E\tilde{\rho}^{\mathrm{B}} = (1/\tau_{\mathrm{B}'})\left[\tilde{\rho}^{\mathrm{B}'}(\mathrm{col}) - \tilde{\rho}^{\mathrm{B}'}\right] \tag{6-206}$$

for B′. Matrix elements of $\tilde{\rho}$(col) are obtained following the procedures of

Chapter V for $\tilde{\rho}^{AB}(\text{col})$ as

$$\left[\tilde{\rho}^{AB}(\text{col})\right]_{ab,\, a'b'} = \sum_{c} \tilde{\rho}^{AB}_{ac,\, a'c}\tilde{\rho}^{B'}_{b,\, b'} \tag{6-207}$$

and for $\tilde{\rho}^{B}(\text{col})$ as

$$\begin{aligned}\left[\tilde{\rho}^{B'}(\text{col})\right]_{c,\, c'} &= \sum_{a} \tilde{\rho}^{AB}_{ac,\, ac'} \sum_{b} \tilde{\rho}^{AB}_{b,\, b} \\ &= \sum_{a} \tilde{\rho}^{AB}_{ac,\, ac'},\end{aligned} \tag{6-208}$$

since

$$\text{Tr}\, \tilde{\rho} = 1.$$

We shall use here the linearization procedure used before, whereby to order of 10^{-5} or less

$$\tilde{\rho}^{AB}_{\text{diag}} \times \tilde{\rho}^{B'}_{\text{off diag}} \sim (1/N_{AB})\tilde{\rho}^{B'}_{\text{off diag}}. \tag{6-209}$$

Labeling the product functions

ϕ^{AB}_{ab}	ab	$\phi^{B'}_{c}$	c
$\alpha\alpha$	1	α	5
$\alpha\beta$	2	β	6
$\beta\alpha$	3		
$\beta\beta$	4		

,

we see that

$$\left[\tilde{\rho}^{AB}(\text{col})\right]_{1,2} = \tilde{\rho}^{AB}_{1,1}\tilde{\rho}^{B'}_{5,6} + \tilde{\rho}^{AB}_{2,2}\tilde{\rho}^{B'}_{5,6}, \tag{6-210}$$

which on linearization,

$$\tilde{\rho}^{AB}_{i,i} = \tfrac{1}{4},$$

becomes

$$\left[\tilde{\rho}^{AB}(\text{col})\right]_{1,2} = \tfrac{1}{2}\tilde{\rho}^{B'}_{5,6}. \tag{6-211}$$

In the same way we obtain

$$\left[\tilde{\rho}^{B'}(\text{col})\right]_{5,6} = \tilde{\rho}^{AB}_{1,2} + \tilde{\rho}^{B'}_{3,4}. \tag{6-212}$$

The NMR absorption is given as

$$\text{Abs} = -\,(\text{AB})\,\text{Im}\left[\tilde{\rho}^{AB}_{1,2} + \tilde{\rho}^{AB}_{1,3} + \tilde{\rho}^{AB}_{2,4} + \tilde{\rho}^{AB}_{3,4}\right] - (\text{B}')\,\text{Im}\,\tilde{\rho}^{B'}_{5,6}. \tag{6-213}$$

Substituting the linearized form of (6-207) and (6-208) into (6-4) one obtains, for example, the $\langle 1\|2\rangle$ matrix element as [note we are using the random field relaxation operator (6-99)]

$$\left[i\left(\omega - \omega_{0\mathrm{B}} - \frac{J_{\mathrm{AB}}}{2}\right) - \frac{1}{T_{\mathrm{lA}}} - \frac{1}{T_{\mathrm{lB}}} - \frac{1}{T_{\mathrm{tB}}} - \frac{1}{\tau_{\mathrm{AB}}}\right]\tilde{\rho}_{1,2}^{\mathrm{AB}} + \frac{i\,J_{\mathrm{AB}}}{2}\tilde{\rho}_{1,3}^{\mathrm{AB}} + \frac{1}{T_{\mathrm{lA}}}\tilde{\rho}_{3,4}^{\mathrm{AB}} + \frac{1}{2\tau_{\mathrm{AB}}}\tilde{\rho}_{5,6}^{\mathrm{B}'} = \frac{i\beta'}{4}, \tag{6-214}$$

where T_{t} and T_{l} stand for transverse and longitudinal relaxation, respectively. The density matrix equation for the exchanging system appears as

$$\begin{bmatrix} \\ \\ i\omega\mathcal{G} + A \\ \\ \\ \end{bmatrix}\begin{bmatrix} \tilde{\rho}_{1,2}^{\mathrm{AB}} \\ \tilde{\rho}_{1,3}^{\mathrm{AB}} \\ \tilde{\rho}_{2,4}^{\mathrm{AB}} \\ \tilde{\rho}_{3,4}^{\mathrm{AB}} \\ \tilde{\rho}_{5,6}^{\mathrm{B}'} \end{bmatrix} = \frac{i\beta'}{4}\begin{bmatrix} 1 \\ 1 \\ 1 \\ 1 \\ 2 \end{bmatrix}. \tag{6-215}$$

The contribution to the A matrix containing just the time constants, T's and τ's is given as

$$\begin{bmatrix}
\begin{pmatrix} -(1/T_{\mathrm{lA}}) - (1/T_{\mathrm{lB}}) \\ -(1/T_{\mathrm{tB}}) - 1/\tau_{\mathrm{AB}} \end{pmatrix} & 0 & 0 & 1/T_{\mathrm{lA}} & 1/2\tau_{\mathrm{AB}} \\
0 & \begin{pmatrix} -(1/T_{\mathrm{lA}}) - (1/T_{\mathrm{tA}}) \\ -(1/T_{\mathrm{lB}}) - 1/2\tau_{\mathrm{AB}} \end{pmatrix} & (1/T_{\mathrm{lB}}) + 1/2\tau_{\mathrm{AB}} & 0 & 0 \\
0 & (1/2\tau_{\mathrm{AB}}) + 1/T_{\mathrm{lB}} & \begin{pmatrix} -(1/T_{\mathrm{lA}}) - (1/T_{\mathrm{tA}}) \\ -(1/T_{\mathrm{lB}}) - 1/2\tau_{\mathrm{AB}} \end{pmatrix} & 0 & 0 \\
1/T_{\mathrm{lA}} & 0 & 0 & \begin{pmatrix} -(1/T_{\mathrm{lA}}) - (1/T_{\mathrm{lB}}) \\ -(1/T_{\mathrm{tB}}) - 1/\tau_{\mathrm{AB}} \end{pmatrix} & 1/2\tau_{\mathrm{AB}} \\
1/\tau_{\mathrm{B}'} & 0 & 0 & 1/\tau_{\mathrm{B}'} & \begin{pmatrix} -(1/T_{\mathrm{lB}}) - (1/T_{\mathrm{tB}'}) \\ -1/\tau_{\mathrm{B}'} \end{pmatrix}
\end{bmatrix} \tag{6-216}$$

4. Relaxation[33]

Up to now in this chapter, for simplicity, the effects of nuclear spin relaxation on the NMR lineshape have been simulated by adding phenomenological linewidth parameters $1/T$ to the diagonal elements of the coefficient matrix of the density matrix equations (except for the last example). Thus, each line comes with a characteristic width $1/T$. Strictly speaking, these terms only account for relaxation due to viscosity, field inhomogeneity, and other instrumental effects. In fact, they have been used quite indiscriminately as a catch-all for all relaxation effects.

The errors in doing this are not serious for systems containing only $\frac{1}{2}$ spins with the same "longish" relaxation times (0.5 sec and larger). However, consider, for example, a two spin molecule with scalar coupling relaxing via the random field mechanism. If one of the spins has short relaxation times, the spins become decoupled from one another. Such an effect could not be simulated with phenomenological linewidth parameters in the diagonal elements of the coefficient matrix.

In fact, operators R_m have been derived in Chapter IV which account for the contributions from different relaxation mechanisms. Therefore, in obtaining the density matrix equations for an exchanging system,

$$\dot{\tilde{\rho}} = i[\tilde{\rho}, \mathcal{H}] + R\rho + E\rho, \tag{6-217}$$

each interaction which contributes to relaxation, when indicated, should be taken into account by writing

$$R\rho = \sum_m R_m \rho, \tag{6-218}$$

where m sums over different mechanisms, see Chapter IV.

5. Quadrupolar Nuclei

a. Introduction

The principal contribution to NMR relaxation of nuclei of $I > \frac{1}{2}$ comes from nuclear electric quadrupole induced transitions. To calculate NMR lineshapes for exchanging systems containing quadrupolar nuclei it is necessary to take explicit account of quadrupolar relaxation by use of the appropriate operator, see Chapter IV, Section 2a.

Molecular rotation and reorientation processes produce fluctuations in the interactions between the nuclear electric quadrupole moment, e^2qQ, and the z component of the electric field gradient around the quadrupolar nucleus. Fluctuating magnetic fields resulting from this interaction drive transitions among the nuclear moments of spins with $I > \frac{1}{2}$ bringing about relaxation.

Quadrupole relaxation in a molecule has two effects on the NMR lineshape. It can broaden the resonance of the quadrupolar nucleus and if the latter is scalar coupled to other nuclei in the same molecular species, quadrupolar relaxation can bring about decoupling to a degree which depends upon T_{1q} and the coupling constants. Since T_{1q} becomes smaller with decreasing temperature, decoupling will be more pronounced at lower temperatures.

We shall first consider NMR lineshapes in systems containing one or more quadrupolar nuclei in the absence of any exchange process on the NMR time scale. Later, exchange effects will be included.

b. Quadrupole Relaxation Times from NMR Lineshapes

Relaxation times of quadrupolar nuclei are traditionally measured with pulse or saturation techniques. In addition, quadrupole induced relaxation may be measured indirectly from the NMR lineshape of a nucleus, $I = \frac{1}{2}$, scalar coupled to the quadrupolar nucleus of interest. Thus, fast quadrupole relaxation will average the scalar coupling between the two nuclei. Examples of systems where NMR lineshape changes are governed by quadrupole relaxation effects follow below.

In the basal ^{11}B NMR lineshape of pentaborane ^{10}B becomes decoupled from ^{11}B due to quadrupole relaxation effects.[34]

Proton NMR of methyls in methyl nitrate[35] and trimethyl alkylammonium[36] salts are triplets due to $J(^{14}N,H)$. Reducing the temperature caused line averaging due to nitrogen quadrupole relaxation. NMR lineshapes were analyzed to give T_{1q}.

The proton resonance of borazole in organic solvents shows both one bond ^{11}B,H and ^{14}N,H coupling.[37] Decreasing the temperature below $-8°C$ causes the fine structure to average due to ^{11}B and ^{14}N quadrupole relaxation. Thus T_{1q} for ^{11}B and ^{14}N was measured using the proton NMR lineshapes. From these results, together with the assumption that τ_{qN} and τ_{qB} were the same, it was found that the z component of the electric field gradient at ^{11}B and ^{14}N was also the same. Also, using the Gierer Wirtz modification of the Stokes equation

$$\tau_x = r_{solu} V\eta / r_{solv}, \qquad (6\text{-}219)$$

where r_{solu} and r_{solv} denote radii of solute and solvent molecules, V is the hard sphere volume of the solute, and η is the viscosity, the authors estimated quadrupole coupling constants of 7.6 ± 2.9, 3.6 ± 1.3, and 1.4 ± 0.5 MHz for ^{10}B, ^{11}B, and ^{15}N in borazole.

We shall now derive density matrix equations to describe the NMR lineshape of a typical system subject to quadrupole relaxation effects. Such a system could be the molecule HD where H,$I = \frac{1}{2}$ is coupled to D,$I = 1$.

The NMR spectrum of HD consists of an equal doublet for the deuterium part and a triplet for the hydrogen, the line separations of 40 Hz[38] being due to J(H,D). Quadrupole induced transitions among the deuterium states causes the proton transitions to mix and if deuterium relaxation is fast enough will average the proton triplet into a single line. The density matrix equations (low power),

$$O = i\left[\tilde{\rho}, \overline{\mathcal{H}}\right] + R_q\tilde{\rho}, \tag{6-220}$$

where R_q is the quadrupole relaxation operator, are obtained as follows.

Let the product functions of HD be listed 1–6,

H	D	label	H	D	label
α	+	1	β	+	4
α	0	2	β	0	5
α	−	3	β	−	6

(6-221)

The density matrix elements required to obtain the proton absorbtion will be $\tilde{\rho}_{1,4}$, $\tilde{\rho}_{2,5}$, and $\tilde{\rho}_{3,6}$, while to obtain the deuterium absorption we need $\tilde{\rho}_{1,2}$, $\tilde{\rho}_{2,3}$, $\tilde{\rho}_{4,5}$, and $\tilde{\rho}_{5,6}$. The deuterium NMR absorption is given by

$$\text{Abs}(\omega) = -\sqrt{2}\ \text{Im}(\tilde{\rho}_{1,2} + \tilde{\rho}_{2,3} + \tilde{\rho}_{4,5} + \tilde{\rho}_{5,6}), \tag{6-222}$$

and the proton absorbtion by

$$\text{Abs}_{\text{H}}(\omega) = -\text{Im}(\tilde{\rho}_{1,4} + \tilde{\rho}_{2,5} + \tilde{\rho}_{3,6}). \tag{6-223}$$

Since the deuterium resonance consists of two degenerate pairs, we can use equivalence factoring by writing

$$\tilde{\rho}_{1,2} + \tilde{\rho}_{2,3} = \tilde{\rho}_{\text{I}}, \qquad \tilde{\rho}_{4,5} + \tilde{\rho}_{5,6} = \tilde{\rho}_{\text{II}}, \tag{6-224}$$

and thus reduce the number of equations needed to plot the deuterium resonance to two. The defining parameters are identified as $\omega_{0\text{H}}$ and $\omega_{0\text{D}}$ (chemical shifts), $\omega_{1\text{H}}$ and $\omega_{1\text{D}}$ rf power used to observe the H and D resonances, respectively, and J, the H,D coupling constant. The Hamiltonian is given (weak coupling) as

$$\overline{\mathcal{H}}_{\text{HD}} = (\omega_{0\text{H}} - \omega)I_{\text{H}}^z + (\omega_{0\text{D}} - \omega)I_{\text{D}}^z + JI_{\text{H}}\cdot I_{\text{D}} - \omega_{1\text{H}}I_{\text{H}}^x + \omega_{1\text{D}}I_{\text{D}}^x. \tag{6-225}$$

Substituting (6-225) into (6-220), using R_q at the intermediate narrowing level, see (6-139) to (6-142), and taking matrix elements listed in (6-222) and (6-223) generates the coupled density matrix equations. The relaxation terms derived at intermediate narrowing can be abbreviated using the definitions

$$\mathcal{J}(\alpha\omega_{0i}) = \frac{2\tau_q}{1 + (\alpha\omega_{0i})^2\tau_q^2} \tag{6-226}$$

and

$$A = \frac{1}{4I(2I-1)}\left(\frac{e^2qQ}{\hbar}\right). \tag{6-227}$$

The coupled density matrix equations are then displayed in matrix form, using these definitions as shown in (6-228) and (6-229).

$$\begin{bmatrix} \begin{pmatrix} i\left(\omega - \omega_{0D} - \dfrac{J}{2}\right) \\ -\dfrac{A^2}{10}[9\mathcal{J}(0) + 15\mathcal{J}(\omega_{0D}) + 6\mathcal{J}(2\omega_{0D})] \end{pmatrix} & 0 \\ 0 & \begin{pmatrix} i\left(\omega - \omega_{0D} + \dfrac{J}{2}\right) \\ -\dfrac{A^2}{10}[9\mathcal{J}(0) + 15\mathcal{J}(\omega_{0D}) + 6\mathcal{J}(2\omega_{0D})] \end{pmatrix} \end{bmatrix} \begin{bmatrix} \tilde{\rho}_{1,2} + \tilde{\rho}_{2,3} \\ \tilde{\rho}_{4,5} + \tilde{\rho}_{5,6} \end{bmatrix} = \frac{i\hbar\omega_{0D}\mathscr{H}_{1D}}{6kT} \begin{bmatrix} \sqrt{2} \\ \sqrt{2} \end{bmatrix} \qquad (6\text{-}228)$$

$$\begin{bmatrix} \begin{pmatrix} i(\omega - \omega_{0H} - J) \\ -\dfrac{A^2}{10}[12\mathcal{J}(\omega_{0D}) + 6\mathcal{J}(2\omega_{0D})] \end{pmatrix} & \dfrac{A^2}{10}6\mathcal{J}(2\omega_{0D}) & \dfrac{A^2}{10}12\mathcal{J}(\omega_{0D}) \\ \dfrac{A^2}{10}6\mathcal{J}(2\omega_{0D}) & \begin{pmatrix} i(\omega - \omega_{0H}) \\ -\dfrac{A^2}{10}12\mathcal{J}(2\omega_{0D}) \end{pmatrix} & \dfrac{A^2}{10}6\mathcal{J}(2\omega_{0D}) \\ \dfrac{A^2}{10}12\mathcal{J}(\omega_{0D}) & \dfrac{A^2}{10}6\mathcal{J}(2\omega_{0D}) & \begin{pmatrix} i(\omega - \omega_{0H} + J) \\ -\dfrac{A^2}{10}[12\mathcal{J}(\omega_{0D}) + 6\mathcal{J}(2\omega_{0D})] \end{pmatrix} \end{bmatrix} \begin{bmatrix} \tilde{\rho}_{1,4} \\ \tilde{\rho}_{2,5} \\ \tilde{\rho}_{3,6} \end{bmatrix} = \frac{i\hbar\omega_{0H}\mathscr{H}_{1H}}{6kT} \begin{bmatrix} \frac{1}{2} \\ \frac{1}{2} \\ \frac{1}{2} \end{bmatrix} \qquad (6\text{-}229)$$

Inspection of the deuterium coefficient matrix (6-228) shows how the deuterium linewidth

$$(A^2/10)(9\mathcal{J}(0) + 15\mathcal{J}(\omega_{0D}) + 6\mathcal{J}(2\omega_{0D})) \tag{6-230}$$

is indeed $1/T_{2D}$ [see (6-148).] Further, one can see how off diagonal relaxation terms in the proton coefficient matrix (6-229) can bring about signal averaging of proton resonance. Although these relaxation terms look different, there is only one relaxation parameter needed to fit this lineshape and that is

$$\tfrac{1}{16}(e^2qQ/\hbar)^2\tau_q. \tag{6-231}$$

Most commonly, these lineshape calculations are carried out using the extreme narrowing approximation, so then

$$\mathcal{J}(0) = \mathcal{J}(\omega_{0i}) = \mathcal{J}(2\omega_{0i}) = 2\tau_q \tag{6-232}$$

and $1/T_{1q}$ is given by

$$1/T_{1q} = 1/T_{2q} = \tfrac{3}{8}(e^2qQ/\hbar)^2\tau_q = 6A^2\tau_q. \tag{6-233}$$

Thus, we see how when a nucleus with $I > \frac{1}{2}$ in a molecule is coupled with another one of $I = \frac{1}{2}$, one can extract quadrupolar relaxation times from the NMR lineshape for the spin $I = \frac{1}{2}$ resonance. In this way, NMR lineshape methods have been used to obtain quadrupolar relaxation times in numerous systems.[35–37, 39, 40]

c. Rate Processes on the NMR Time Scale

Blackborrow studied the proton NMR lineshape of the adduct of N,N-dimethylaniline with BCl_3, **6**.[41] The N-methyl proton resonance consists of a quartet at 305°K due to ^{11}B, ^{1}H coupling of 2.9 Hz. At higher temperatures, the fine structure

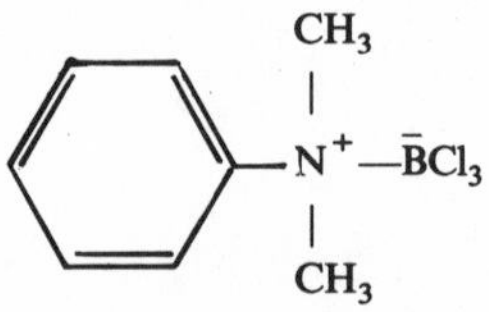

6

is averaged due to boron nitrogen bond exchange, while below 305°K there is line narrowing due to nitrogen quadrupole relaxation. The proton NMR lineshape was used to measure T_{1N}, as well as to elucidate the mechanism of exchange. A dissociation mechanism involving two kinds of ion pairs was proposed

$$\mathrm{DBCl_3} \rightleftharpoons \left(\overset{\delta+}{\mathrm{DBCl_3}} \ldots \overset{\delta-}{\mathrm{Cl}} \right) \rightleftharpoons \mathrm{D_2}\overset{+}{\mathrm{B}}\mathrm{Lc_2} + \mathrm{BCl_4^-},$$

where D stands for the N,N-dimethylaniline moiety

d. Rate Processes on the Molecular Correlation Time Scale

Until now, the exchanging systems discussed involved conditions

$$\tau_c \ll \tau_e,$$

where the molecular correlation is much faster than exchange. However, around $\tau_e \simeq 10\tau_c$ a modification of the previous theory must be used.

The derivation of the density matrix equation describing the dynamics of chemically exchanging systems makes use of the assumption (see Chapter V) that between exchange collisions the density matrix equation obeyed by a molecule A is given as

$$\dot{\tilde{\rho}}^A = -i\left[\mathcal{H}^A, \tilde{\rho}^A\right] + R^A\tilde{\rho}^A, \qquad (6\text{-}235)$$

where R^A is the time-independent relaxation operator. One solves (6-235) as

$$\tilde{\rho}(t - t_0) = e^{(-i\mathcal{L}_s^A + R)(t - t_0)}\rho(\mathrm{col}, t_0), \qquad (6\text{-}236)$$

where one defines

$$-i\left[\mathcal{H}^A, \tilde{\rho}^A\right] = -\mathcal{L}^A\tilde{\rho}^A. \qquad (6\text{-}237)$$

Next, the exchange averaged density matrix operator is weighted by the exchange time as

$$\tilde{\rho}^A = \int_{-\infty}^{t} \frac{e^{-(t-t_0)/\tau_c}}{\tau_e}\tilde{\rho}^A(t - t_0)\, dt_0, \qquad (6\text{-}238)$$

which upon differentiating, see (6-15) and (6-16), becomes

$$\dot{\tilde{\rho}}^A = -i\left[\mathcal{H}^A, \tilde{\rho}^A\right] + R\tilde{\rho}^A - (1/\tau_e)(\tilde{\rho}^A - \rho^A(\mathrm{col})). \qquad (6\text{-}239)$$

If one reanalyzes the situation when the exchange term τ_e approaches the molecular correlation time τ_c, one must redefine R to account for the finiteness of time between collisions. The result is that R (infinite time) must be replaced by

$$R(1 - e^{-(t/\tau_c)}). \qquad (6\text{-}240)$$

Equation (6-240) follows from the form of (4-221) by assuming the extreme narrowing limit and replacing the limit ∞ by t. For intermediate narrowing R will be replaced by

$$\sum R_\alpha\left[1-e^{-(t/\tau_c)}\cos\alpha\omega_0 t+e^{-(t/\tau_c)}\sin\alpha\omega_0 t\right]$$

Thus, for $t>\tau_c$ we obtain the previous relaxation operator valid for infinite times. When the more general form of the relaxation operator (in the extreme narrowing limit) is substituted into (6-239), integration gives

$$\tilde{\rho}^{A}(t-t_0)=\exp\left[\left(-i\bar{\mathfrak{L}}_s+R\right)(t-t_0)-\tau_c Re^{-(t-t_0)/\tau_c}\right]\rho(\text{col},t_0), \qquad (6\text{-}241)$$

and (6-239) is now replaced by

$$\dot{\tilde{\rho}}^{A} = -i\left[\bar{\mathcal{H}}^{A},\tilde{\rho}^{A}\right] + R\tilde{\rho}^{A} + \frac{1}{\tau_{ex}}\left(\rho^{A}(\text{col}) - \tilde{\rho}^{A}\right)$$

$$-R\int_{-\infty}^{t}\frac{e^{-[(t-t_0)/\tau_{ex}]}}{\tau_{ex}}e^{-[(t-t_0)/\tau_c]}\tilde{\rho}^{A}(t-t_0)\,dt. \qquad (6\text{-}242)$$

For $\tau_c<\tau_{ex}$ the dominant contribution from the last term will be

$$-(\tau_c/\tau_{ex})R(\tilde{\rho}^{A}(\text{col}), \qquad (6\text{-}243)$$

which allows us to write (6-242) as

$$\dot{\tilde{\rho}}^{A} = -i\left[\bar{\mathcal{H}}^{A},\tilde{\rho}^{A}\right] + R\tilde{\rho}^{A} + \frac{1}{\tau_{ex}}\left(\tilde{\rho}^{A}(\text{col}) - \tilde{\rho}^{A}\right)$$

$$-(\tau_c/\tau_{ex})R(\tilde{\rho}^{A}(\text{col})). \qquad (6\text{-}244)$$

In the limit of $\tau_c \ll \tau_{ex}$ the last term can be dropped to obtain the previous result.

Examples of systems where quadrupole correlation and exchange are reported to lie on the same time scale include exchange process among molecular complexes of 1,3,5-trinitrobenzene,[42] as well as the interaction of 2-propanol-2D,P, with horse liver dehydrogenate,[43] DH

$$\text{P} + \text{DH} \rightleftarrows \text{PDH}. \qquad (6\text{-}245)$$

In the latter study, Drakenberg used the proton NMR lineshape of isopropanol to obtain the deuterium T_1 values. The derived τ_qs were assumed, in an approximate treatment, to result from the averaging of $1/\tau_q$ and $1/\tau_{ex}$ for isopropanol exchange.[44, 45] However, the above treatment was not used.

6. Effect of Intermediates on NMR Lineshapes[46, 47]

In principle, the NMR lineshape for an exchanging system has contributions from all species present (as long as they contain the isotope whose resonance is being scanned). While intermediates of low concentration and short lifetime give no contribution to the observed linewidth, species of low concentration and intermediate lifetime do contribute to the lineshape. The following hypothetical example illustrates this point.

Consider the first order exchanging system of three species, A, B, and C, each spin $I = \frac{1}{2}$, which interconvert according to the scheme

$$\mathrm{A} \underset{k_{-1}}{\overset{k_1}{\rightleftarrows}} \mathrm{B} \underset{k_{-2}}{\overset{k_2}{\rightleftarrows}} \mathrm{C}. \tag{6-246}$$

While A and C are in abundant concentrations, B is dilute. We set the equilibrium constant

$$K = \frac{(\mathrm{C})}{(\mathrm{A})} = \frac{k_1 k_{-2}}{k_{-1} k_2} = 1 \tag{6-247}$$

equal to one with

$$k_1 = k_{-2} \quad \text{and} \quad k_{-1} = k_2. \tag{6-248}$$

For convenience, the density matrix equations are given as

$$[i\mathscr{I}(\omega - \omega_0) + \mathrm{A}]\rho_{\mathrm{col}} = (ih\omega_0\,\mathscr{H}_1/4kT_{\mathrm{b}})\mathrm{B}_{\mathrm{col}}, \tag{6-249}$$

where ω_0 is the center of gravity of the spectrum under conditions of slow exchange, $\mathrm{B}_{\mathrm{col}}$ is a column vector with all "ones", and $\mathrm{A}\rho_{\mathrm{col}}$ is given as

$$\begin{bmatrix} \begin{pmatrix} i(\omega_0 - \omega_{0\mathrm{A}}) \\ -k_1 - 1/T_{\mathrm{A}} \end{pmatrix} & k_1 & 0 \\ k_{-1} & \begin{pmatrix} i(\omega_0 - \omega_{0\mathrm{B}}) \\ -k_{-1} - k_2 - 1/T_{\mathrm{B}} \end{pmatrix} & k_2 \\ 0 & k_{-2} & \begin{pmatrix} i(\omega_0 - \omega_{0\mathrm{C}}) \\ -k_{-2} - 1/T_{\mathrm{C}} \end{pmatrix} \end{bmatrix} \begin{bmatrix} \tilde{\rho}_{\mathrm{A}} \\ \tilde{\rho}_{\mathrm{B}} \\ \tilde{\rho}_{\mathrm{C}} \end{bmatrix} \tag{6-250}$$

The absorption is

$$\mathrm{Abs} = \mathscr{C}\,\mathrm{Im}[(\mathrm{A})\tilde{\rho}_{\mathrm{A}} + (\mathrm{B})\tilde{\rho}_{\mathrm{B}} + (\mathrm{C})\tilde{\rho}_{\mathrm{C}}]. \tag{6-251}$$

Calculations of the NMR lineshape were carried out for different sets of rate constants consistent with Eqs. (6-247) and (6-248) are shown in Fig. 6-8 for (A) = (C) = 10(B). The unsymmetrical character of these lineshapes arises from the low concentration intermediate at $\nu = 16$ Hz.

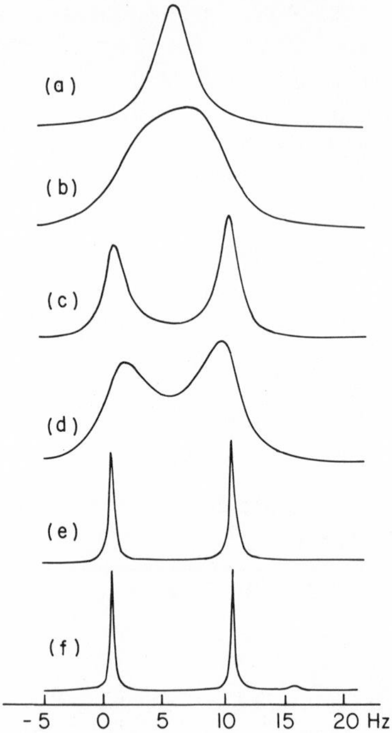

Fig. 6-8. NMR lineshapes for the system A ⇄ B ⇄ C, three $\frac{1}{2}$ spin sites with (A) = (C) = 10(B), $\nu_A = 1$, $\nu_B = 16$, and $\nu_C = 11$ Hz, respectively, $T_A = T_B = T_C = 1$ sec.; values of k_1, k_{-1}, k_2, and k_{-2} in $\sec^{-1}$ are (a) 100, 1000, 1000, 100; (b) 45, 450, 450, 45; (c) 25, 250, 250, 25; (d) 10, 100, 100, 10 (e) 1, 10, 10, 1; (f) 0.1, 1, 1, 0.1.

7. NMR Lineshapes of Solutes in the Nematic Phase

Hitherto in the theory for NMR of molecules we have assumed no preferred orientation. This must be modified for solutes in the nematic phase which do have an orientational preference.[48–51] The necessary assumption for the derivation of the relaxation operator R (Chapter IV) was that the spin–bath interaction was such that

$$\mathrm{Tr_b}\ \mathcal{H}_{sb}\rho_0(\mathcal{H}_b) = 0. \qquad (6\text{-}252)$$

where

$$\mathcal{H}_{sb} = \sum_{\alpha} \nu^{\alpha} \mathcal{G}^{\alpha}, \qquad (6\text{-}253)$$

and ν^{α} and $\mathcal{G}^{\alpha}$, the spin and bath operators, respectively, are defined in Chapter IV.

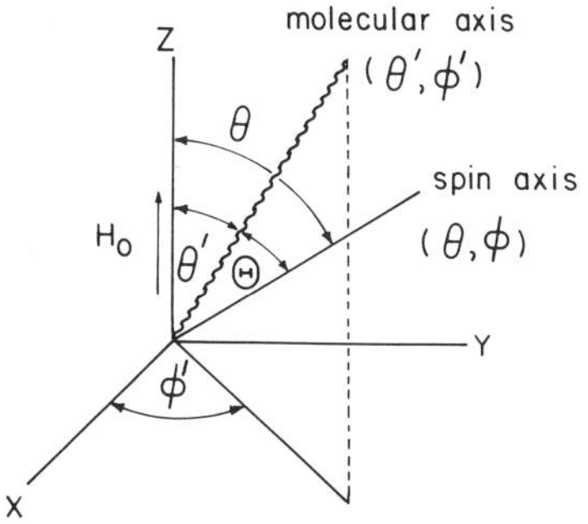

Fig. 6-9. The axis of the molecular coordinate system of the solute species in the nematic phase is defined as (θ', ϕ') relative to the ordering direction (also direction of applied field). The vector $r_{i,j}$ connecting the two spins on the solute molecule has direction (θ, ϕ) relative to the ordering direction and has angle Θ relative to the molecular axis.

Equation (6-252) for the normal liquid state is equivalent to the "classical" requirement that for all "α"

$$\bar{\nu}^{\alpha} = 0 = (1/4\pi)\int\int \nu^{\alpha}(\theta, \phi)\sin\theta \, d\theta \, d\phi. \tag{6-254}$$

The correctness of (6-254) for the dipole-dipole interaction can be readily checked by the reader. For the liquid crystal (nematic phase) we have the situation illustrated in Fig. 6-9. (Note we are assuming the solute molecule has axial symmetry.)

The magnetic field is in the z direction which we also take as the liquid crystal ordering direction. The probability of the axial molecule pointing in the (θ', ϕ') direction is given by $f(\theta')$.† The classical representation of Eq. (6-252) for a liquid crystal will be given as

$$\mathrm{Tr_b}\,\mathcal{H}_{\mathrm{sb}}\rho_0(\mathcal{H}_0) = \sum_{\alpha} \mathcal{G}^{\alpha}\bar{\nu}^{\alpha}, \tag{6-255}$$

where

$$\bar{\nu}^{\alpha} = (1/4\pi)\int_0^{2\pi}\int_0^{\pi}\int_0^{2\pi} f(\theta')\nu^{\alpha}(\theta, \phi)\sin\theta' \, d\phi' \, d\theta' \, d\omega. \tag{6-256}$$

The angle ω is shown in Fig. 6-10.†
From spherical trigonometry, we have

$$\cos\theta = \cos\theta' \cos\Theta + \sin\theta' \sin\Theta \cos\omega. \tag{6-257}$$

$\bar{\nu}^{\alpha}$ for a liquid crystal will be zero for all except

$$\begin{aligned} \nu_{i,j}^{0}(\theta, \phi) &= KP_2(\theta), \qquad K = -2\hbar^2\gamma_i\gamma_j/r_{i,j}^3, \\ P_2(\theta) &= \tfrac{1}{2}(3\cos^2\theta - 1), \end{aligned} \tag{6-258}$$

†The distribution function $f(\theta')$ is given by $f(\theta') = e^{-(u_2/kT)}/\int e^{-(u_2/kT)}\sin\theta' \, d\theta'$, where $u_2 = \mathrm{const}\,P_2(\theta')$ and P_n is an nth order Legendre polynomial.

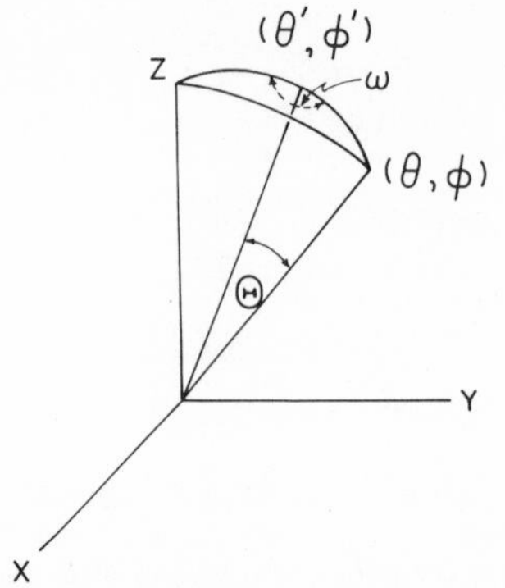

Fig. 6-10. Same as Fig (6-9) with the addition that ω is the angle between the planes defined by the pairs of lines {ordering direction, (θ', ϕ')} and {(θ', ϕ'), (θ, ϕ)}.

where i and j are a pair of interacting nuclear spins and $r_{i,j}$ the distance between them. Substituting (6-258) into (6-256)

$$\frac{1}{4\pi}\int_0^{2\pi}\int_0^{\pi} d\phi' \sin\theta' \, d\theta' \frac{1}{2\pi}\int_0^{2\pi} d\omega \, f(\theta') \frac{K}{2}$$

$$\times\Big[3(\cos^2\theta' \cos^2\Theta + 2\cos\theta' \cos\Theta \sin\theta' \sin\Theta \cos\omega$$

$$+\sin^2\theta' \sin^2\Theta \cos^2\omega) - 1\Big] = \bar{\nu}^0(\theta) \tag{6-259}$$

which, after performing the ω integration, becomes

$$\bar{\nu}^0_{\mathrm{lc}}(\theta) = \nu^0(\Theta)\tfrac{1}{2}\int_0^{\pi} P_2(\theta') f(\theta') \sin\theta' \, d\theta', \tag{6-260}$$

where the subscript lc means liquid crystal. Thus, one has that

$$\mathrm{Tr}\ \mathcal{H}_{\mathrm{sb}}\rho_0(\mathcal{H}_0) = \mathcal{G}^0\bar{\nu}^0_{\mathrm{lc}} = \overline{\mathcal{H}}_{\mathrm{sb}}. \tag{6-261}$$

To calculate NMR lineshapes for solutes in the nematic phase, one must include the result $\overline{\mathcal{H}}_{\mathrm{sb}}$ in the spin Hamiltonian to give

$$\overline{\mathcal{H}}_{\mathrm{lc}} = \sum_i (\omega_{0i} - \omega) I^z + \sum_{i>j} J_{i,j} I_i \cdot I_j + \sum_{i>j} \mathcal{G}^0_{i,j}\bar{\nu}^0_{i,j} + \sum_i \omega_i I_i^x. \tag{6-262}$$

Note that the extra term in $\bar{\nu}^0$ comes summed over all dipolar interactions between pairs of nuclear spins.

By use of $\overline{\mathcal{H}}_{\mathrm{lc}}$ in the density matrix equation it is possible to calculate the NMR spectrum of a solute in the nematic phase. Note that since $\bar{\nu}^0$ contains both internuclear distance(s) and the angles, the spectrum depends on the molecular structure. In fact, this technique has been used to determine the structures of numerous molecules as solutes in the nematic phase.[50]

The density matrix equation at low power for liquid crystal solutes

$$\dot{\tilde{\rho}} = 0 = i\left[\tilde{\rho}, \overline{\mathcal{H}}_{\mathrm{lc}}\right] + E\tilde{\rho} + R\tilde{\rho} \tag{6-263}$$

can also be used to calculate lineshapes for exchanging systems in the nematic phase. Evaluation of the exchange term $E\tilde{\rho}$ follows the procedures given in Chapters V and VI. One can use a phenomenological linewidth parameter instead of $R\tilde{\rho}$. However, the relaxation operators differ from those used for isotropic media and have to be evaluated separately.†

For solutes in the nematic phase, the previous "necessary assumption" (6-252) must be now replaced by

$$\mathrm{Tr_b}\ \mathcal{H}_{\mathrm{sb}}^{\mathrm{eff}}\rho_0(\mathcal{H}_0) = 0, \tag{6-264}$$

where eff stands for effective and

$$\begin{aligned}\mathcal{H}_{\mathrm{sb}}^{\mathrm{eff}} &= \mathcal{H}_{\mathrm{sb}} - \overline{\mathcal{H}}_{\mathrm{sb}} = \sum_{\alpha} \nu_{\mathrm{eff}}^{\alpha}\mathcal{G}^{\alpha} \\ &= \sum_{\alpha} \mathcal{G}^{\alpha}\nu^{\alpha}(\theta, \phi) - \mathcal{G}^{0}\bar{\nu}^{0}.\end{aligned} \tag{6-265}$$

Thus, the three bath operators which apply are

$$\begin{aligned}\nu_{\mathrm{eff}}^{0} &= \nu^{0}(\theta, \phi) - \nu^{0}(\Theta)\tfrac{1}{2}\int_0^{\pi} P_2(\theta')f(\theta')\sin\theta'\, d\theta', \\ \nu_{\mathrm{eff}}^{\pm 1} &= \nu^{\pm 1}(\theta, \phi), \qquad \nu_{\mathrm{eff}}^{\pm 2} = \nu^{\pm 2}(\theta, \phi),\end{aligned} \tag{6-266}$$

where the $\nu^{\alpha}(\theta,\phi)$ are, for dipole-dipole relaxation given in (4-153)–(4-156). The spin operators $\mathcal{G}_{i,j}^{\alpha}$ are as listed in (4-150)–(4-152). Further derivation of the nematic phase dipolar relaxation operator follows the methods of Chapter IV with the proviso that the angular evaluations given before as

$${}_{\alpha}c_{(i,j)}^{+}(\tau) = (1/4\pi)\int_0^{2\pi}\int_0^{\pi} \nu_{i,j}^{\alpha}(0)\nu_{i,j}^{-\alpha}(0)\sin\theta\, d\theta\, d\phi\, e^{-(\tau/\tau_d)\ddagger} \tag{4-158b}$$

must now, using $\nu_{\mathrm{eff}}^{\alpha}$ instead of ν^{α}, be weighted by

$$(1/4\pi)\int_0^{2\pi}\int_0^{\pi} d\phi' \sin\theta'\, d\theta'\ (1/2\pi)\int_0^{2\pi} d\omega\, f(\theta'). \tag{6-267}$$

Thus, in the nematic phase,

$$c_{\alpha\ (i,j)}^{+}(\tau) = (1/4\pi)\int_0^{2\pi}\int_0^{\pi} \nu_{\mathrm{eff}}^{\alpha}(0)\nu_{\mathrm{eff}}^{-\alpha}(0)\, d\phi' \sin\theta'\, d\theta' (1/2\pi)\int_0^{2\pi} d\omega f(\theta')g(\tau), \tag{6-268}$$

†In the discussion which follows on relaxation in the nematic phase, it should be kept in mind that in practice there are significant temperature gradients in nematic samples. These gradients entirely determine the observed linewidths. Should techniques be developed to eliminate thermal gradients, then the other relaxation mechanisms discussed here become important.

‡Here τ_d stands for dipolar correlation time.

where $g(\tau)$ is a correlation function which, in general, will not be a simple exponential. Note that in the liquid crystal, the temperature dependence of the relaxation operator arises from both $f(\theta')$ and $g(\tau)$. The treatment for other relaxation mechanisms follows that given here for dipole-dipole relaxation.

This concludes the discussion of NMR lineshapes of solutes in the nematic phase. Using the above procedure one can take account of exchange processes as well as relaxation mechanisms. Extensions can be made to conditions of high rf field and double resonance using the results of this section together with information in Chapters VII and VIII. For nonaxial molecules the calculations are more complicated.

8. Experimental Problems

The experimental problems encountered in obtaining a good NMR spectrum of a system not involved in an exchange process are enormously compounded when one turns to NMR lineshapes for exchanging systems obtained as a function of temperature. Then, the results become extremely sensitive to the external environment. Some of these more important questions will now be discussed.

To control the temperature of the NMR sample, it is customary to blow thermostatted gas through the insert. Because there is not much room in the insert, temperature gradients across the sample of 2–3° often ensue. This undesirable situation can be alleviated with a fast flow of gas. As of this writing, a properly thermostatted insert is not commercially available. However, some improved designs have recently been described.[52]

NMR parameters—shifts, coupling constants, and relaxation times can all vary with temperature. For this reason, NMR spectra have to be obtained over a wide temperature range in which exchange is too slow to affect the NMR lineshape. The NMR parameters obtained from these spectra are plotted as a function of temperature. As these plots tend to be largely linear, extrapolation into the temperature regions where exchange influences the lineshape gives estimates of the NMR parameters at the higher temperatures. Linewidth parameters can be obtained in this fashion also. Effects due to viscosity can be taken into account by using the known variation of apparent linewidth with viscosity.

Finally, to improve the fit of theoretical to experimental lineshapes, it is sometimes necessary to iterate the parameters.

The comparison of NMR lineshapes was handled originally with linewidth measurements or peak to valley intensity ratios. A plot was constructed from calculations of some parameter characteristic of the lineshape versus rate constant. Then, reading the experimental parameter off the plot gave the rate constant.

More recently, lineshape comparisons have been carried out with a point by point least squares fit.

The accuracy with which rate constants can be evaluated depends on the nature of the NMR parameters being averaged. A system in which several parameters of quite different magnitudes become averaged by exchange is more likely to give NMR lineshapes which are sensitive to changes in rate constant over a wide range of rates than one where only a single parameter undergoes averaging.

Furthermore, as shown above, strong coupling and extensive fine structure leads to the extrication of far more extensive kinetic information than is possible with first order systems. Against this must be balanced the errors introduced in fitting a spectrum with many NMR parameters. Also, there is the problem that the accuracy with which a rate constant can be obtained may vary with its magnitude and this must be taken into account when Eyring and Arrhenius plots are evaluated for activation parameters.

9. Summary

The theory for low rf field NMR lineshapes in chemically exchanging systems has been described at different levels of approximation starting with the coupled two site problem, multisite half-spin exchange, weak coupling, and finally exchange among strongly coupled systems. The connection between weakly coupled systems and their Kubo–Anderson–Sack simulation is made. Also, simplifications due to spin equivalence are explained. Density matrix equations are derived for several examples of exchanging systems.

Effects are treated due to reaction intermediates, quadrupole relaxation, and solutes in the nematic phase. Finally, there are some remarks about experimental problems.

Problems

1. Find the weak coupling absorption peak strengths relative to a spin $\frac{1}{2}$ absorption for the Hamiltonian

$$\mathcal{H} = \omega_{0\mathrm{H}} \sum_{i=1}^{3} I_i^z + \omega_{0\mathrm{B}} \sum_{j=1}^{4} I_j^z + \sum_{i<j} J_{i,j} I_i \cdot I_j.$$

A group of three spins shift $\omega_{0\mathrm{H}}$ couple equivalently to a group of four spin shift $\omega_{0\mathrm{B}}$.

2. Calculate the low power density matrix equations for the system of $\frac{1}{2}$ spins undergoing exchange:

$$\mathcal{H}_{AB} = \omega_{0A} I_A^Z + \omega_{0B} I_B^Z, \qquad \mathcal{H}_C = \omega_{0C} I_C^Z$$

$$AB \rightleftarrows BA, \qquad AB + C \rightleftarrows AC + B$$

3. Calculate the low power density matrix equations for six equivalent spins separately exchanging with a single spin.

4. Calculate the temperature dependence of the observed chemical shift of a spin with axially symmetric chemical shifts located on a solute molecule dissolved in a liquid crystal in the nematic phase.

REFERENCES

1. J. I. Kaplan and G. Fraenkel, *J. Amer. Chem. Soc.* **94**, 2907 (1972).
2. H. S. Gutowsky, D. M. McCall, and C. P. Slichter, *J. Chem. Phys.* **21**, 279 (1953).
3. H. S. Gutowsky and A. Saika, *J. Chem. Phys.* **21**, 1688 (1953).
4. H. S. Gutowsky and C. H. Holm, *J. Chem. Phys.* **25**, 1288 (1956).
5. H. M. McConnell and C. H. Holm, *J. Chem. Phys.* **28**, 430 (1958).
6. P. W. Anderson, *J. Phys. Soc. Japan* **9**, 316 (1954).
7. R. Kubo, *J. Phys. Soc. Japan.* **9**, 935 (1954).
8. R. Kubo, *Nuovo Cimento, Suppl.* **6**, 1063 (1957).
9. R. A. Sack, *Mol. Phys.* **1**, 163 (1958).
10. M. Saunders, *Tetrahedron Lett.* 1699 (1963).
11. L. M. Reeves and K. M. Shaw, *Can J. Chem.* **48**, 3641 (1970).
12. J. I. Kaplan, *J. Magn. Reson.*, **21**, 153 (1976).
13. I. Solomon and W. Bloembergen, *J. Chem. Phys.* **25**, 261 (1958).
14. J. P. Maher and D. F. Evans, *J. Chem. Soc.* 5534 (1963).
15. J. Soulati, K. L. Henhold, and J. P. Oliver, *J. Amer. Chem. Soc.* **73**, 5694 (1971).
16. K. L. Henhold, J. Soulati, and J. P. Oliver, *J. Amer. Chem. Soc.* **91**, 3171 (1969).
17. R. M. Lynden-Bell, *Prog. Nucl. Magn. Reson. Spectros.* **2**, 163 (1967).
18. A. Loewenstein and T. M. Connor, *Ber. Bunsenges. Phys. Chem.* **67**, 280 (1963).
19. L. W. Reeves, *Advan. Phys. Org. Chem.* **5**, 187 (1967).
20. C. S. Johnson, *Advan. Magn. Reson.* **1**, 33 (1965).
21. G. Binsch, *Top. Stereochem.* **3**, 97 (1968).
22. G. Binsch, *Top. Stereochem.* **3**, 1304 (1968).
23. L. M. Jackman and F. A. Cotton (eds.), "Dynamic Nuclear Magnetic Resonance Spectroscopy," Chaps. 6–12. Academic Press, New York, 1975.
24. J. I. Kaplan, *J. Chem. Phys.* **28**, 278 (1958).
25. J. I. Kaplan, *J. Chem. Phys.* **28**, 462 (1958).
26. S. Alexander, *J. Chem. Phys.* **37**, 967, 974 (1962).
27. S. Alexander, *J. Chem. Phys.* **38**, 1787 (1963).
28. S. Alexander, *J. Chem. Phys.* **40**, 2741 (1964).
29. G. Fraenkel and D. T. Dix, *J. Amer. Chem. Soc.* **88**, 979 (1966).
30. J. A. Pople, W. A. Schneider, and H. J. Bernstein, "High Resolution Nuclear Magnetic Resonance." McGraw-Hill, New York, 1959.
31. G. Fraenkel, C. E. Cottrell, and D. T. Dix, *J. Amer. Chem. Soc.* **93**, 1704 (1971).
32. S. Q. A. Rizvi, Ohio State Univ., Columbus, Ohio, private communication.

33. N. C. Pyper, *Mol. Phys.* **20**, 449 (1971).
34. D. W. Lowman, P. D. Ellis, and J. D. Odam, *J. Magn. Reson.* **8**, 289 (1972).
35. C. Sheppard, T. Schaeffer, B. W. Boodwin, and J. Rao, *Can. J. Chem.* **49**, 5138 (1971).
36. D. W. Larsen, *J. Phys. Chem.* **75**, 509 (1971).
37. G. M. Whitesides, L. Regen, J. B. Lisle, and R. Mays, *J. Chem. Phys.* **76**, 2871 (1972).
38. H. Y. Carr and E. M. Purcell, *Phys. Rev.* **88**, 415 (1952).
39. J. A. Kintzinger, J. M. Lehnard, and R. L. Williams, *Mol. Phys.* **17**, 135 (1969).
40. Ch. Brevard, J. P. Kintzinger, and J. M. Lehn, *Tetrahedron* **28**, 2429, 2477 (1972).
41. J. R. Blackborrow, *J. Chem. Soc. A* 2143 (1973).
42. Ch. Brevard and J. M. Lehn, *J. Amer. Chem. Soc.* **92**, 4987 (1970).
43. T. Drakenberg, *J. Magn. Reson.* **15**, 354 (1974).
44. J. E. Anderson and P. A. Fryer, *J. Chem. Phys.* **50**, 3784 (1969).
45. A. C. Marshall, *J. Chem. Phys.* **52**, 2527 (1970).
46. J. J. Led and D. M. Grant, *J. Amer. Chem. Soc.* **99**, 5845 (1977).
47. S. O. Chan and L. W. Reeves, *J. Amer. Chem. Soc.* **95**, 670 (1973).
48. A. Saupe and G. Englert, *Phys. Rev. Letters* **11**, 462 (1963).
49. G. Englert and A. Saupe, *Z. Naturforsch.* **19a**, 172 (1964).
50. P. Diehl and C. L. Khetrapal, *in* "NMR Basic Principles and Progress," (P. Diehl, E. Fluck and R. Kosfeld, eds.), p. 1. Springer-Verlag, New York, 1969.
51. P. G. de Gennes, "The Physics of Liquid Crystals." Clarenden Press, Oxford, 1974.
52. D. A. Gilles, *in* "Nuclear Magnetic Resonance," (R. K. Harris, ed.), Specialist Periodical Reports, Series B, p. 156. The Chemical Society, London, 1974.

Chapter VII
NMR OF EXCHANGING SYSTEMS AT HIGH rf FIELDS

Commonly, NMR lineshapes for exchanging systems are obtained under conditions of slow sweep and low rf field. The theory which covers these experiments has been described in Chapter VI. Now, however, we show how one can lift the low rf field approximation and calculate NMR lineshapes for exchanging systems as a function of the rf field.

An example of a high rf lineshape calculation will be given. It is shown that high rf field spectra are more sensitive to slow exchange than low power spectra.

More importantly, the results obtained with the methods of this chapter constitute the starting point in the theory for double resonance lineshapes of exchanging systems.

In this chapter we will discuss the calculation of the NMR absorption of an exchanging system under conditions of high rf power.[1, 2] The principal difference in calculating low power compared to the high power absorption is that, in the latter case, one needs to include *all elements* of the density matrix equation

$$\dot{\tilde{\rho}} = [\tilde{\rho}, \mathcal{H}] + R\tilde{\rho} + E\tilde{\rho} \tag{7-1}$$

instead of just those $(\dot{\tilde{\rho}})_{i,k}$ where $m_i - m_k = +1$ needed for low power. Thus, for a two $\frac{1}{2}$ spin system AB, with product functions $\alpha\alpha$ (1), $\alpha\beta$ (2), $\beta\alpha$ (3), and $\beta\beta$ (4) at low power, one needs just four elements—$(\dot{\tilde{\rho}})_{1,2}$, $(\dot{\tilde{\rho}})_{1,3}$, $(\dot{\tilde{\rho}})_{2,4}$, and $(\dot{\tilde{\rho}})_{3,4}$, whereas at high power we require all 16 elements, the four just mentioned and twelve additional ones—$(\dot{\tilde{\rho}})_{i,j}$ correctly neglected at low power

i, k	i, k	i, k	i, k
1, 2	2, 4	1, 4	1, 1
2, 1	4, 2	4, 1	2, 2
1, 3	3, 4	2, 3	3, 3
3, 1	4, 3	3, 2	4, 4

At high power diagonal elements of the density matrix do not have their equilibrium values and have to be solved for. In the computer calculation of the absorption one can still reduce the dimension of the matrix to be inverted as will be shown shortly.

As previously discussed, we use the high power form of the density matrix, expressed as

$$\tilde{\rho} = (1/N) + \rho_1 , \tag{7-2}$$

where N is the dimensionality of the spin space (number of states) for the species in question.

Let us consider again, as in Chapter VI, the bimolecular exchange process

$$\mathrm{AB + B' \rightleftharpoons AB' + B}. \tag{6-200}$$

The density matrix equation, its exchange, and intermolecular relaxation contributions are described in Chapter VI, Section 3. As before, the absorption is given by

$$\text{Abs} = -(\text{AB}) \operatorname{Im}(\tilde{\rho}_{1,2}^{\text{AB}} + \tilde{\rho}_{1,3}^{\text{AB}} + \tilde{\rho}_{2,4}^{\text{AB}} + \tilde{\rho}_{3,4}^{\text{AB}}) - (\text{B}') \operatorname{Im} \tilde{\rho}_{5,6}^{\text{B}'} \cdot \tag{6-213}$$

For comparison with the low rf power condition described in Chapter VI here is the $\langle 1|\ |2\rangle$ element for high power,

$$\begin{aligned}(\dot{\tilde{\rho}}^{\text{AB}})_{1,2} = 0 = i\Big[(E_2 - E_1)\tilde{\rho}_{1,2}^{\text{AB}} + \frac{\omega_1}{2}(\tilde{\rho}_{1,4}^{\text{AB}} + \tilde{\rho}_{1,1}^{\text{AB}} - \tilde{\rho}_{3,2}^{\text{AB}} \\ - \tilde{\rho}_{2,2}) + J\tilde{\rho}_{1,2}^{\text{AB}}\Big] + \frac{1}{\tau_{\text{AB}}}\big[\tilde{\rho}_{5,6}^{\text{B}'}(\tilde{\rho}_{1,1}^{\text{AB}} + \tilde{\rho}_{2,2}^{\text{AB}}) - \tilde{\rho}_{1,2}^{\text{AB}}\big] \\ - \tilde{\rho}_{1,2}^{\text{AB}}\left[\frac{1}{T_{1\text{A}}} + \frac{1}{T_{1\text{B}}} + \frac{1}{T_{\text{tB}}}\right] + \tilde{\rho}_{3,4}^{\text{AB}}\frac{1}{T_{1\text{A}}},\end{aligned} \tag{7-3}$$

where the E_i terms are diagonal elements of $\mathcal{H}^0$ [(7-3)].

$$\begin{aligned}E_1 &= \tfrac{1}{2}(\omega_{0\text{A}} - \omega) + \tfrac{1}{2}(\omega_{0\text{B}} - \omega) + \tfrac{1}{2}J, \\ E_2 &= \tfrac{1}{2}(\omega_{0\text{A}} - \omega) - \tfrac{1}{2}(\omega_{0\text{B}} - \omega) - \tfrac{1}{2}J, \\ E_3 &= -\tfrac{1}{2}(\omega_{0\text{A}} - \omega) + \tfrac{1}{2}(\omega_{0\text{B}} - \omega) - \tfrac{1}{2}J, \\ E_4 &= -\tfrac{1}{2}(\omega_{0\text{A}} - \omega) - \tfrac{1}{2}(\omega_{0\text{B}} - \omega) + \tfrac{1}{2}J.\end{aligned} \tag{7-4}$$

As seen from (7-2), the diagonal matrix elements are written as

$$\tilde{\rho}_{1,1}^{AB} = (1/N_{AB}) + \delta\tilde{\rho}_{1,1}^{AB}, \qquad \tilde{\rho}_{5,5}^{B'} = (1/N_{B'}) + \delta\tilde{\rho}_{5,5}^{B'}, \tag{7-5}$$

where N_{AB} and $N_{B'}$ are 4 and 2, respectively.

Inspection of (7-3) reveals that in this set of coupled equations the exchange contributions contain terms quadratic in the matrix elements which must be solved for to ultimately give the NMR absorption. It is considerably more complicated and time consuming to solve coupled inhomogeneous second order equations than the linear sets needed for the low power NMR. However, a linearization procedure has been developed (see Chapter V) in which certain bilinear terms may be safely dropped. This is called selective neglect of bilinear terms, SNOB. We now proceed to show how this approximation applies to the different exchange terms in the coupled equations for (7-2).

The exchange contributions in the density matrix equations contain terms such as

$$\tilde{\rho}_{i,j}^{AB}\tilde{\rho}_{k,l}^{B'} \sim (\tilde{\rho}_{\text{off diag}})^2, \qquad i \neq j, \quad k \neq l,$$

$$\tilde{\rho}_{i,i}^{AB}\tilde{\rho}_{k,l}^{B'} = \left[(1/\mathrm{N}_{AB}) + \delta\tilde{\rho}_{i,i}^{AB}\right]\tilde{\rho}_{k,l}^{B'}, \tag{7-6}$$

$$\tilde{\rho}_{i,i}^{AB}\tilde{\rho}_{k,k}^{B'} = (1/\mathrm{N}_{AB}\mathrm{N}_{B'}) + (\delta\rho_{i,i}^{AB}/\mathrm{N}_{B}) + (\delta\rho_{k,k}^{AB}/\mathrm{N}_{AB}) + \delta\rho_{i,i}^{AB}\delta\rho_{k,k}^{B'}.$$

The maximum value matrix elements $\tilde{\rho}_{i,j}$ and $\delta\rho_{k,k}$ can have is $\sim 10^{-6}$. Thus, in a sum consisting of first and second order terms the latter can be safely dropped, being smaller than the former by a factor of 10^{-6} or less. Thus, bilinear terms of the type

$$\tilde{\rho}_{i,j}\delta\tilde{\rho}_{k,k}, \quad \delta\tilde{\rho}_{i,i}\delta\tilde{\rho}_{k,k}, \quad \text{and} \quad \tilde{\rho}_{i,j}\tilde{\rho}_{k,l} \tag{7-7}$$

are neglected where they appear. In this way, the exchange term in (7-3) becomes

$$(1/\tau_{AB})\left[\tfrac{1}{2}\tilde{\rho}_{5,6}^{B'} - \tilde{\rho}_{1,2}^{AB}\right], \tag{7-8}$$

and so (7-3) is rewritten as

$$0 = i\left[(E_2 - E_1)\tilde{\rho}_{1,2}^{AB} + (\omega_1/2)(\tilde{\rho}_{1,4}^{AB} + \delta\rho_{1,1}^{AB} - \tilde{\rho}_{3,2}^{AB} - \delta\tilde{\rho}_{2,2}^{AB}) + J_{AB}\tilde{\rho}_{1,3}^{AB}\right]$$
$$-\tilde{\rho}_{1,2}^{AB}\left[(1/T_{1A}) + (1/T_{1B}) + (1/T_{tB})\right] + \tilde{\rho}_{3,3}^{AB}(1/T_{1A}). \tag{7-9}$$

All $\Delta m = \pm 1$ equations will be linearized in similar fashion. Next, consider the equations which do not appear using the low power approximations, for instance, that for $\langle 2|\dot{\tilde{\rho}}^{AB}|2\rangle$, which is given as

$$\dot{\tilde{\rho}}^{AB}_{2,2} = 0 = i\left[\frac{\hbar\omega_1}{2}\left(\tilde{\rho}^{AB}_{2,4} - \tilde{\rho}^{AB}_{4,2} + \tilde{\rho}^{AB}_{2,1} - \tilde{\rho}^{AB}_{1,2}\right) + \frac{J}{2}\left(\tilde{\rho}^{AB}_{2,3} - \tilde{\rho}^{AB}_{3,2}\right)\right] + \frac{1}{\tau_{AB}}\left[\tilde{\rho}^{B'}_{6,6}\left(\tilde{\rho}^{AB}_{1,1} + \tilde{\rho}^{AB}_{2,2}\right) - \tilde{\rho}^{AB}_{2,2}\right] - \tilde{\rho}^{AB}_{2,2}\left(\frac{1}{T_{1A}} + \frac{1}{T_{1B}}\right) + \tilde{\rho}^{AB}_{4,4}\frac{1}{T_{1A}} + \tilde{\rho}^{AB}_{1,1}\frac{1}{T_{1B}} - \frac{\hbar\omega_0}{4kT_b}\left(\frac{1}{T_{1A}} + \frac{1}{T_{1B}}\right). \tag{7-10}$$

On linearization the exchange term (in τ_{AB}) becomes

$$(1/\tau_{AB})\left[\tfrac{1}{2}\delta\rho_{6,6} + \tfrac{1}{2}(\delta\rho_{1,1} - \delta\rho_{2,2})\right]. \tag{7-11}$$

Entirely new terms appearing at high power are the $\Delta m = 0$ equation $\langle 2|\dot{\tilde{\rho}}^{AB}|3\rangle$ and the double quantum equations $\Delta m = \pm 2$, one of which is for $\langle 1|\dot{\tilde{\rho}}^{AB}|4\rangle$. These two equations, just mentioned, are given as

$$\dot{\tilde{\rho}}_{2,3} = 0 = i\left[(E_3 - E_2)\tilde{\rho}^{AB}_{2,3} + \frac{\hbar\omega_1}{2}\left(\tilde{\rho}^{AB}_{2,1} - \tilde{\rho}^{AB}_{4,3} + \tilde{\rho}^{AB}_{2,4} - \tilde{\rho}^{AB}_{1,3}\right) + \frac{J}{2}\left(\tilde{\rho}^{AB}_{2,2} - \tilde{\rho}^{AB}_{3,3}\right)\right] + \frac{1}{\tau_{AB}}\left(\tilde{\rho}^{AB}_{1,3}\tilde{\rho}^{B'}_{6,5} + \tilde{\rho}^{AB}_{2,4}\tilde{\rho}^{B'}_{6,5} - \tilde{\rho}^{AB}_{2,3}\right) - \tilde{\rho}^{AB}_{2,3}\left(\frac{1}{T_{1A}} + \frac{1}{T_{1B}} + \frac{1}{T_{tA}} + \frac{1}{T_{tB}}\right), \tag{7-12}$$

$$\dot{\tilde{\rho}}_{1,4} = 0 = i\left[(E_4 - E_1)\tilde{\rho}^{AB}_{1,4} + \frac{\hbar\omega_1}{2}\left(\tilde{\rho}^{AB}_{1,2} - \tilde{\rho}^{AB}_{3,4} + \tilde{\rho}^{AB}_{1,3} - \tilde{\rho}^{AB}_{2,4}\right)\right] + \frac{1}{\tau_{AB}}\left(\tilde{\rho}^{AB}_{1,3}\tilde{\rho}^{B'}_{5,6} + \tilde{\rho}^{AB}_{2,4}\tilde{\rho}^{B'}_{5,6} - \tilde{\rho}^{AB}_{1,4}\right) - \tilde{\rho}^{AB}_{1,4}\left(\frac{1}{T_{1A}} + \frac{1}{T_{tA}} + \frac{1}{T_{1B}} + \frac{1}{T_{tB}}\right). \tag{7-13}$$

In the last two equations linearization consists of dropping the terms which are nonlinear in the off diagonal elements. Thus, all that remains in the exchange term of (7-12) is

$$-\,(1/\tau_{AB})\tilde{\rho}^{AB}_{2,3} \tag{7-14}$$

and similarly for (7-13) we are left with

$$-\,(1/\tau_{AB})\tilde{\rho}^{AB}_{1,4}\,. \tag{7-15}$$

Explicit calculations including these dropped bilinear terms have shown that to order 10^{-6} for all values of τ_{AB} there is no observable change in the NMR absorption compared to that obtained with the SNOB approximation.[2] Linearization as seen in (7-11) is obviously correct as we are dropping terms which are at least 10^{-6} smaller than those we keep (i.e., we keep $\frac{1}{2}\tilde{\rho}_{3,2}^{AB}$ but drop $\tilde{\rho}_{5,6}^{B'}(\delta\rho_{1,1} + \delta\rho_{2,2})$. On the other hand, in (7-12) and (7-13) we neglect the entire $\tilde{\rho}$(col) term in both equations. To see the effect of this approximation, rewrite (7-13) by solving for $\tilde{\rho}_{1,4}^{AB}$ and lump all the relaxation terms into one $1/T_{\text{eff}}$ (eff = effective).

$$\tilde{\rho}_{1,4}^{AB} = \frac{(\omega_1/2)(\tilde{\rho}_{1,2}^{AB} - \tilde{\rho}_{3,4}^{AB} + \tilde{\rho}_{1,3}^{AB} - \tilde{\rho}_{2,4}^{AB}) + (1/\tau_{AB})(\tilde{\rho}_{1,3}^{AB}\tilde{\rho}_{5,6}^{B'} + \tilde{\rho}_{2,4}^{AB}\tilde{\rho}_{5,6}^{B'})}{-i(E_4 - E_1) + (1/T_{\text{eff}}) + 1/\tau_{AB}}. \tag{7-16}$$

It would appear that for $\tau_{AB} \sim 10^{-6}$ sec the two kinds of terms in the numerator would be of the same order of magnitude. This is true but unimportant for at even relatively slow exchange rates the τ_{AB}^{-1} in the denominator broadens the line beyond experimental detection. Thus, in the very fast exchange limit nonlinear terms may well contribute to an enormously broadened line but are not numerically or experimentally important.

At this point it is worth considering the significance of the double quantum transition. First, substitute for $E_4 - E_1$, in (7-13), assume slow exchange (i.e., leave out $1/\tau_{AB}$), and one obtains

$$\tilde{\rho}_{1,4}^{AB} = \frac{(i\omega_1/2)(\tilde{\rho}_{1,2}^{AB} - \tilde{\rho}_{3,4}^{AB} + \tilde{\rho}_{1,3}^{AB} - \tilde{\rho}_{2,4}^{AB})}{-i(2\omega - \omega_{0B} - \omega_{0A}) + 1/T_{\text{eff}}}. \tag{7-17}$$

Therefore, there will be a resonance when

$$\omega = \tfrac{1}{2}(\omega_{0A} + \omega_{0B}), \tag{7-18}$$

exactly in the middle of the AB spectrum. The magnitude of the absorption will be second order in ω_1, as we note that the $\tilde{\rho}^{AB}$ elements in the numerator of (7-17) are already first order in ω_1. The $\tilde{\rho}_{1,4}^{AB}$ element, itself, does not directly contribute to the magnetization, Eq. (6-213), but indirectly by contributing to $\tilde{\rho}_{1,2}^{AB}$ and other $\Delta m = \pm 1$ terms.

The full set of 20 equations can be formally written as

$$\begin{bmatrix} i\mathcal{I}\omega + D & R \\ L & K \end{bmatrix} \begin{bmatrix} \rho_{\text{I}} \\ \rho_{\text{II}} \end{bmatrix} = \begin{bmatrix} 0 \\ B_{\text{II}} \end{bmatrix}, \tag{7-19}$$

where ρ_{I}, dimension 12, contains all the $\Delta m = \pm 1, \pm 2$ transitions. The latter two equations ($\dot{\tilde{\rho}}_{1,4}^{AB}$ and $\dot{\tilde{\rho}}_{4,1}^{AB}$) have to be scaled by $\frac{1}{2}$ to obtain the form given in (7-19). Next, expand (7-19) to give

$$(i\mathcal{I}\omega + D)\rho_{\text{I}} + R\rho_{\text{II}} = 0, \tag{7-20}$$

$$L\rho_{\text{I}} + K\rho_{\text{II}} = B_{\text{II}}. \tag{7-21}$$

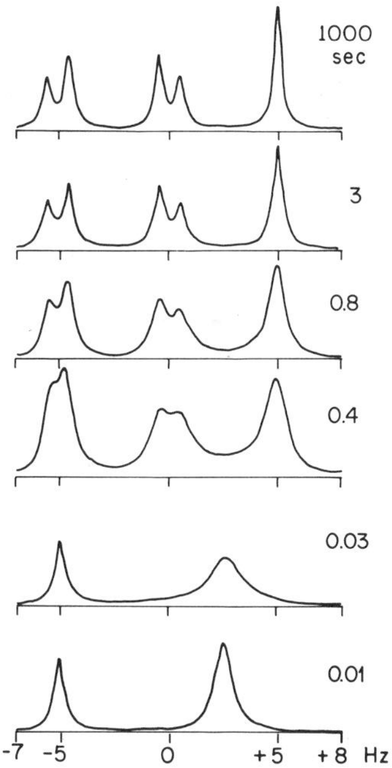

Fig. 7-1. NMR lineshapes for exchanging system (6-200), (AB) = (B′) $\nu_A = -5$, ν_B (of AB) = 0, $\nu_{B'} = 5$, $J(AB) = 1$, Hz, respectively, all T_1 values 10 sec., all T_2 values 0.8 sec., $\nu_1 = 0.02$ Hz, τ_{AB} values on figure.

Solving the coupled equations (7-20) and (7-21) for ρ_I leads to

$$(i\mathcal{G}\omega + D - RK^{-1}L)\rho_I = -RK^{-1}B_{II} \tag{7-22}$$

and ρ_I is obtained according to the methods described in Chapter III.

Examples of NMR lineshapes calculated, using the SNOB approximation, at different rf fields for several values of τ_{AB} are illustrated in Figs. 7-1 and 7-2. The shifts and relaxation times used in these calculations are listed in the legend. Mainly, these lineshapes show the expected broadening of the B and B′ resonances. At low rates of exchange $\tau = 3$–1000 sec. The high rf lineshapes are considerably more sensitive to changes in the rate of exchange than at limiting low rf field. Also, the double quantum transition makes its expected weak appearance at the center of the AB spectrum. This absorption is extremely sensitive to exchange rate and disappears altogether at moderate rates of exchange.

One can see that NMR of exchanging systems at high rf field has some advantages over low power NMR.

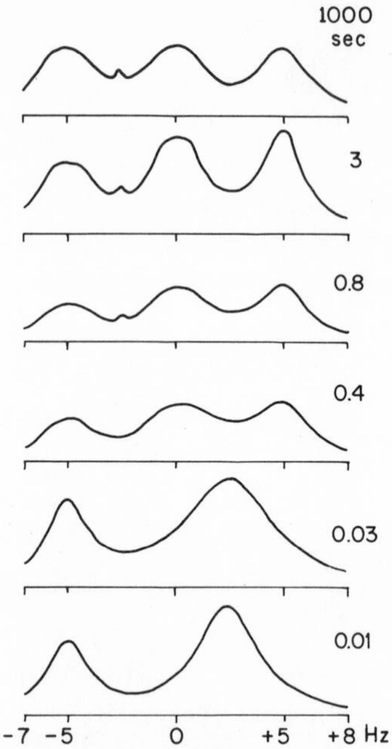

Fig. 7-2. As in Fig. 7-1 but with $\not{\nu}_1$ at 0.5 Hz.

Another application of high power NMR lineshapes is based on the fact that it enables one to distinguish between intra- and intermolecular dipole-dipole relaxation. Harris and Pyper observed that whereas at low power the NMR lineshapes for an AB system (spin $\frac{1}{2}$) were identical for these two kinds of relaxation, under saturation conditions the lineshapes looked quite different. The experimental results for 2-chloroacrylonitrile fitted best for intermolecular relaxation.[3] Also, calculations of high rf field NMR lineshapes for 2, 6-difluoro-3, 4, 5-trichloropyridine[4–6] and 2-bromothiazole[7] have been reported.

Problem

1. Derive the density matrix equations for the system

$$AH + BH^* \rightleftarrows AH^* + BH$$

at high power. Show that this system can be represented by Bloch type equations. Note A and B contain no nuclear spins but represent environments.

REFERENCES

1. J. I. Kaplan, *J. Chem. Phys.* **28**, 278 (1958).
2. G. Fraenkel, J. I. Kaplan, and P. P. Yang, *J. Chem. Phys.* **60**, 2574 (1974).
3. R. K. Harris and K. M. Worville, *J. Magn. Reson.* **9**, 394 (1973).
4. R. K. Harris, N. C. Pyper, R. E. Richards, and G. W. Schultz, *Mol. Phys.* **19**, 145 (1970).
5. R. K. Harris and N. C. Pyper, *Mol. Phys.* **20**, 467 (1971).
6. R. K. Harris and K. M. Worville, *J. Magn. Reson.* **9**, 383 (1973).
7. R. K. Harris and N. C. Pyper, *Mol. Phys.* **23**, 277 (1971).

Chapter VIII
DOUBLE RESONANCE

1. Theory of NMDR Lineshapes

In this chapter we show how nuclear magnetic double resonance (NMDR), one high rf and one low rf, lineshape analysis may be used to measure rates of exchange slower than is possible using single resonance NMR. To see this qualitatively consider the now familiar exchanging system

$$AB + B' \rightleftarrows AB' + B, \tag{8-1}$$

where A, B, and B′ are species which contain one proton, each B is identical to B′, and A is coupled to B. The effect of exchange on the low rf single resonance lineshape is to average the shifts of bound B (in AB) with free B as well as the AB coupling. Slow exchange causes the B (in AB) and B′ resonances to broaden slightly. When the free B′ resonance is strongly irradiated at ω_1, the B observed resonance of AB at ω_2 (weak) will broaden additionally due to transfer of saturation by exchange.

In the past, effects in double resonance not involving exchanging systems—multiplet formation, Overhauser effects, and spin decoupling have been used to analyze for NMR parameters.[1–6] The origin of these effects will be brought out during our discussion of double resonance.

The saturation-recovery technique[7–11] pioneered by Hoffmann and Forsén[7, 8] will not be discussed in this chapter (see Chapter 9).

At this point, before solving the appropriate NMDR equation[12, 13] we will briefly indicate how spin decoupling comes about. Spin decoupling is just the reduction of the multiplet structure of the two spins which are coupled via the Hamiltonian

$$\overline{\mathcal{H}}_2 = \overline{\mathcal{H}}_0^{AB} + \not{\omega}_1[I_A^x + I_B^x]. \tag{8-2}$$

Imagine the frequency to be exactly at the B resonance, i.e.,

$$\omega_1 = \omega_{0B}, \tag{8-3}$$

then (8-2) becomes

$$\overline{\mathcal{H}}_2 = (\omega_{0A} - \omega_1)I_A^z + J(I_A^x I_B^x + I_A^y I_B^y + I_A^z I_B^z) + \not{\omega}_1(I_A^x + I_B^x). \tag{8-4}$$

For $\not{\omega}_1/J > 1$ the effective quantization direction of the B spins is seen to be no longer in the z direction but in the x direction. We indicate this by rotating the B system about the y axis so that

$$\begin{aligned} x &\to z', & I_B^x &\to I_B^{z'}, \\ y &\to y', & I_B^y &\to I_B^{y'}, \\ z &\to x', & I_B^z &\to -I_B^{x'}. \end{aligned} \tag{8-5}$$

With this rotation (8-4) becomes

$$\overline{\mathcal{H}}_2 = (\omega_{0A} - \omega_1)I_A^z + J(I_A^x I_B^{z'} + I_A^y I_B^{y'} + I_A^z I_B^{x'}) + \not{\omega}_1(I_A^x + I_B^{z'}). \tag{8-6}$$

The coupling interaction (i.e., the operator multiplying J) thus will no longer contribute to a first order energy correction to the eigenvalues of I_A^z and $I_B^{z'}$.

Proof:

$$I_A^z|m\rangle = m|m\rangle, \qquad I_B^{z'}|n\rangle = n|n\rangle, \tag{8-7}$$

and

$$\langle mn|I_A^x I_B^{z'} + I_A^y I_B^{y'} + I_A^z I_B^{x'}|mn\rangle = 0. \tag{8-8}$$

Thus, the first order shift is eliminated.

We shall develop the theory of steady state double resonance by solving for the lineshapes of our model exchanging system (8-1).[12] However, the method employed here applies to any exchanging system under the influence of two rf fields, one weak, $\not{\omega}_2$ (observe) and the other strong, $\not{\omega}_1$.

$$\begin{aligned} \mathcal{H} &= \sum_i \omega_{0i} I_i^z + \sum_{i>j} J_{i,j} I_i \cdot I_j + \sum_i \not{\omega}_1[I_i^x \cos \omega_1 t + I_i^y \sin \omega_1 t] \\ &\quad + \sum_i \not{\omega}_2[I_i^x \cos \omega_2 t + I_i^y \sin \omega_2 t] \\ &= \mathcal{H}_0 + \mathcal{H}_{rf1} + \mathcal{H}_{rf2}, \end{aligned} \tag{8-9}$$

and in the system rotating at ω_1

$$\overline{\mathcal{H}} = \sum_i (\omega_{0i} - \omega_1) I_i^z + \sum_{i>j} J_{i,j} I_i \cdot I_j + \tfrac{1}{2}\omega_1 \sum_i (I_i^+ + I_j^-)$$
$$+ \omega_2 \sum_i \left(I_i^+ e^{-i(\omega_2-\omega_1)t} + I_i^- e^{+i(\omega_2-\omega_1)t} \right)$$
$$= \overline{\mathcal{H}}_0 + \overline{\mathcal{H}}_{rf1} + \overline{\mathcal{H}}_{rf2} \,. \tag{8-10}$$

The density matrix equations are given as usual as

$$\dot{\rho}^{AB} = -i[\mathcal{H}^{AB}, \rho^{AB}] + R\rho^{AB} + E\rho^{AB} \tag{8-11}$$
$$\dot{\rho}^{B'} = -i[\mathcal{H}^{B'}, \rho^{B'}] + R\rho^{B'} + E\rho^{B'}.$$

We shall use the random field relaxation operator, see Chapter IV. However, to include Overhauser effects it would be necessary to add terms for inter- and *intradipolar* relaxation (see Chapter IV). To simplify the treatment these effects have been neglected.

In the *product representation*, the $E\rho$ exchange elements for (8-1) are given by, see Chapter V,

$$[E\rho^{AB}]_{ab,a'b'} = \frac{1}{\tau_{AB}} \left[\sum_c \rho^{AB}_{ac,a'c} \rho^{B'}_{b,b'} - \rho^{AB}_{ab,a'b'} \right], \tag{8-12}$$

$$[E\rho^{B'}]_{c,c'} = \frac{1}{\tau_{B'}} [\rho^{AB}_{ac,a'c} - \rho^{B'}_{c,c'}], \tag{8-13}$$

where the subscripts are labels of product states.

To solve (8-11) to all orders we must go into the rotating frame. The density matrix equations for the exchanging system (8-1) then become, see Appendix A,

$$\dot{\tilde{\rho}}^{AB} = -i[\overline{\mathcal{H}}_0^{AB}, \tilde{\rho}^{AB}] - i\omega_1[(I_A^x + I_B^x), \tilde{\rho}^{AB}]$$
$$- i\frac{\omega_2}{2}[(I_A^+ + I_B^+)e^{i\Delta\omega t} + (I_A^- + I_B^-)e^{-i\Delta\omega t}, \tilde{\rho}^{AB}]$$
$$+ R(\tilde{\rho}^{AB} - \rho_0^{AB}) + E\tilde{\rho}^{AB}, \tag{8-14}$$
$$\dot{\tilde{\rho}}^{B'} = -i[\overline{\mathcal{H}}_0^{B'}, \tilde{\rho}^{B'}] - i\omega_1[I_{B'}^x, \tilde{\rho}^{B'}]$$
$$- i\frac{\omega_2}{2}[I_{B'}^+ e^{i\Delta\omega t} + I_{B'}^- e^{-i\Delta\omega t}, \tilde{\rho}^{B'}]$$
$$+ R(\tilde{\rho}^{B'} - \rho_0^{B'}) + E\tilde{\rho}^{B'}, \tag{8-15}$$

where $\Delta\omega = \omega_1 - \omega_2$ and

$$\overline{\mathcal{H}}_0^{AB} = (\omega_{0A} - \omega_1)I_A^z + (\omega_{0B} - \omega_1)I_B^z + J I_A \cdot I_B \,, \tag{8-16}$$

$$\overline{\mathcal{H}}_0^{B'} = (\omega_{0B} - \omega_1)I_{B'}^z \,. \tag{8-17}$$

We now write

$$\tilde{\rho}^{AB} = \tilde{\rho}_{sr}^{AB} + \rho_{+}^{AB} e^{i(\omega_2 - \omega_1)t} + \rho_{-}^{AB} e^{-i(\omega_2 - \omega_1)t},$$
$$\tilde{\rho}^{B'} = \tilde{\rho}_{sr}^{B'} + \rho_{+}^{B'} e^{i(\omega_2 - \omega_1)t} + \rho_{-}^{B'} e^{-i(\omega_2 - \omega_1)t}, \tag{8-18}$$

where the $\tilde{\rho}_{sr}$ operators are the single resonance, sr, solutions of (8-14) and (8-15) with $\not{\omega}_2 = 0$, that is for the saturation resonance at frequency ω_1 as described in Chapter VII.

Note also that as $\tilde{\rho}$ is hermitian

$$\rho_{+}^{AB} = (\rho_{-}^{AB})^{\dagger}, \tag{8-19}$$

so we need only obtain the ρ_{+}^{AB} or the ρ_{-}^{AB} elements. To obtain the equation for ρ_{+}^{AB} and $\rho_{+}^{B'}$, we substitute $\tilde{\rho}$ in (8-18) into (8-14) and (8-15) and collect all terms with the time dependence

$$e^{i(\omega_2 - \omega_1)t}$$

with the result that

$$i(\omega_2 - \omega_1)\rho_{+}^{AB} = i\left[\rho_{+}^{AB}, \mathcal{H}_0^{AB}\right] - i\not{\omega}_1\left[I_A^x + I_B^x, \rho_{+}^{AB}\right]$$
$$-\tfrac{1}{2} i\not{\omega}_2\left[I_A^- + I_B^-, \tilde{\rho}_{sr}^{AB}\right] + R\rho_{+}^{AB} + E\rho_{+}^{AB}, \tag{8-20}$$

$$i(\omega_2 - \omega_1)\rho_{+}^{B'} = i\left[\rho_{+}^{B'}, \mathcal{H}_0^{B'}\right] - i\not{\omega}_1\left[I_{B'}^x, \rho_{+}^{B'}\right]$$
$$-\tfrac{1}{2} i\not{\omega}_2\left[I_{B'}^-, \tilde{\rho}_{sr}^{B'}\right] + R\rho_{+}^{B'} + E\rho_{+}^{B'}. \tag{8-21}$$

Note that in the relaxation operator in these ρ_+ equations, ρ_+ is to be used and not $\rho_+ - \rho_0$.

Before going on to evaluate (8-20) and (8-21) in the product representation, let us first calculate the expression for the double resonance absorption. This we obtain by evaluating the x component of magnetization in the laboratory frame. We recall that the absorption at frequency ω_2 will be given by the component of magnetization out of phase with the applied rf field. Thus, the absorption at frequency ω_2 is the coefficient of M_x going as $\sin \omega_2 t$. Now, we have in the laboratory system that

$$M_x = (\mathrm{AB})\ \mathrm{Tr}\ \rho^{AB}(I_A^x + I_B^x) + (\mathrm{B'})\ \mathrm{Tr}\ \rho^{B'} I_{B'}^x. \tag{8-22}$$

Going into the rotating coordinate system at frequency ω_1 (8-22) becomes

$$M_x = (\mathrm{AB})\ \mathrm{Tr}\ \tilde{\rho}^{AB} e^{i\omega_1 t I_{AB}^z} I_{AB}^x e^{-i\omega_1 t I_{AB}^z}$$
$$+ (\mathrm{B'})\ \mathrm{Tr}\ \tilde{\rho}^{B'} e^{i\omega_1 t I_{B'}^z} I_{B'}^x e^{-i\omega_1 t I_{B'}^z}, \tag{8-23}$$

where we abbreviate $I_A^x + I_B^x$ as I_{AB}^x.

Replacing I^x by $\frac{1}{2}(I^+ + I^-)$, using Eq. (8-A12) (see Appendix A) and (8-18), M_x becomes

$$\begin{aligned} M_x = &\tfrac{1}{2}(\mathrm{AB}) \operatorname{Tr} \rho_{\mathrm{sr}}^{\mathrm{AB}}\left(I_{\mathrm{AB}}^+ e^{i\omega_1 t} + I_{\mathrm{AB}}^- e^{-i\omega_1 t}\right) \\ &+ \tfrac{1}{2}(\mathrm{B}') \operatorname{Tr} \rho_{\mathrm{sr}}^{\mathrm{B}'}\left(I_{\mathrm{B}'}^+ e^{i\omega_1 t} + I_{\mathrm{B}'}^- e^{-i\omega_1 t}\right) \\ &+ \tfrac{1}{2}(\mathrm{AB}) \operatorname{Tr}\left\{\rho_+^{\mathrm{AB}} e^{i(\omega_2 - 2\omega_1)t} I_{\mathrm{AB}}^- + \rho_-^{\mathrm{AB}} e^{-i(\omega_2 - 2\omega_1)t} I_{\mathrm{AB}}^+\right\} \\ &+ \tfrac{1}{2}(\mathrm{B}') \operatorname{Tr}\left\{\rho_+^{\mathrm{B}'} e^{i(\omega_2 - 2\omega_1)t} I_{\mathrm{B}'}^- + \rho_-^{\mathrm{B}'} e^{-i(\omega_2 - 2\omega_1)t} I_{\mathrm{B}'}^+\right\} \\ &+ \tfrac{1}{2}(\mathrm{AB}) \operatorname{Tr}\left\{\rho_+^{\mathrm{AB}} e^{i\omega_2 t} I_{\mathrm{AB}}^+ + \rho_-^{\mathrm{AB}} e^{-i\omega_2 t} I_{\mathrm{AB}}^-\right\} \\ &+ \tfrac{1}{2}(\mathrm{B}') \operatorname{Tr}\left\{\rho_+^{\mathrm{B}'} e^{i\omega_2 t} I_{\mathrm{B}'}^+ + \rho_-^{\mathrm{B}'} e^{-i\omega_2 t} I_{\mathrm{B}'}^-\right\}. \end{aligned} \tag{8-24}$$

Collecting all terms going as $\sin \omega_2 t$, we obtain

$$\begin{aligned} \mathrm{Abs}(\omega_2) \simeq &\tfrac{1}{2} i(\mathrm{AB})\left\{\operatorname{Tr} \rho_+^{\mathrm{AB}} I_{\mathrm{AB}}^+ - \operatorname{Tr} \rho_-^{\mathrm{AB}} I_{\mathrm{AB}}^-\right\} \\ &+ \tfrac{1}{2} i(\mathrm{B}')\left\{\operatorname{Tr} \rho_+^{\mathrm{B}'} I_{\mathrm{B}'}^+ - \operatorname{Tr} \rho_-^{\mathrm{B}'} I_{\mathrm{B}'}^-\right\}. \end{aligned} \tag{8-25}$$

Using the product states

AB	B′
$\phi_1 = \alpha\alpha$	
$\phi_2 = \alpha\beta$	$\phi_5 = \alpha$
$\phi_3 = \beta\alpha$	$\phi_6 = \beta$
$\phi_4 = \beta\beta$	

(8-26)

and (8-19) reduces Eq. (8-26) to

$$\begin{aligned} \mathrm{Abs}(\omega_2) \simeq &-(\mathrm{AB}) \operatorname{Im}\left[(\rho_+^{\mathrm{AB}})_{2,1} + (\rho_+^{\mathrm{AB}})_{3,1} + (\rho_+^{\mathrm{AB}})_{4,2} + (\rho_+^{\mathrm{AB}})_{4,3}\right] \\ &-(\mathrm{B}') \operatorname{Im}(\rho_+^{\mathrm{B}'})_{6,5}\,. \end{aligned} \tag{8-27}$$

Before deriving various elements of the ρ_+ equation, (8-20), some remarks are in order about $E\tilde{\rho}_+$. In evaluating the exchange term we can safely accomplish a linearization analogous to the SNOB approximation of Chapter VII. This amounts to keeping terms of the type

$$(\tilde{\rho}_{\mathrm{sr}}^x)_{i,i}(\rho_+^y)_{k,l} \simeq (1/N_x)(\rho^y)_{k,l}, \tag{8-28}$$

while dropping

$$(\tilde{\rho}_{\mathrm{sr}}^x)_{i,j}(\rho_+^y)_{k,k} \quad \text{and} \quad (\tilde{\rho}_s^x)_{i,j}(\rho_+^y)_{k,l}\,. \tag{8-29}$$

The next step is to obtain all matrix elements of the ρ_+ equation. There results a set of 20 coupled equations in the ρ_+ elements which contain also, see below, elements of $\tilde{\rho}_{\mathrm{sr}}$. In a typical experiment a system is irradiated strongly at a fixed frequency ω_1 and swept with ω_2, a weak observing rf field. First, all elements of $\tilde{\rho}_{\mathrm{sr}}$ (single resonance high rf field) are calculated separately using the methods of Chapter VII. The results are substituted into the ρ_+ equations. These are solved, as described previously, for the elements in Eq. (8-27) which are summed to give the absorption. The parameters required for this calculation are shifts, coupling constants, relaxation times, the rf fields ω_1 and ω_2, the irradiation frequency ω_1, and the ω_2 frequencies needed to trace the low rf response.

Now, let us examine the different elements of Eq. (8-20). The expression for the absorption (8-27) suggests we look at the $\langle 2|\ |1\rangle$ element first [see (7-4) for definition of E_is]

$$\begin{aligned} 0 = i\{[E_1 - E_2 - \omega_2 - \omega_1](\rho_+^{AB})_{2,1} - \tfrac{1}{2}J(\rho_+^{AB})_{3,1} \\ + \tfrac{1}{2}\omega_1[-(\rho_+^{AB})_{1,1} + (\rho_+^{AB})_{2,2} + (\rho_+^{AB})_{2,3} - (\rho_+^{AB})_{4,1}] \\ + \tfrac{1}{2}\omega_2[-(\tilde{\rho}_{sr}^{AB})_{1,1} + (\tilde{\rho}_{sr}^{AB})_{2,2} + (\tilde{\rho}_{sr}^{AB})_{2,3}]\} \\ + (1/\tau_{AB})[\tfrac{1}{2}(\rho_+^{B'})_{6,5} - (\rho_+^{AB})_{2,1}] \\ - [(1/T_{1A}) + (1/T_{1B}) + 1/T_{tB}](\rho_+^{AB})_{2,1} + [1/T_{1A}](\rho_+^{AB})_{4,3}. \end{aligned} \tag{8-29a}$$

Equation (8-29a) should be compared to the $\langle 1|\ |2\rangle$ matrix element obtained using the low power single resonance approximation (see Chapter VI) which is

$$\begin{aligned} \dot{\tilde{\rho}}_{1,2} = 0 = i\{[E_2 - E_1]\tilde{\rho}_{1,2}^{AB} - \tfrac{1}{2}J\tilde{\rho}_{1,3}^{AB} + \tfrac{1}{2}\omega_2[\tilde{\rho}_{2,2}^{AB} - \tilde{\rho}_{1,1}^{AB}]\} \\ + (1/\tau_{AB})[\tfrac{1}{2}\tilde{\rho}_{5,6}^{B'} - \tilde{\rho}_{1,2}] \\ - [(1/T_{1A}) + (1/T_{1B}) + 1/T_{tB}]\tilde{\rho}_{1,2}^{AB} + (1/T_{1A})\tilde{\rho}_{3,4}^{AB}. \end{aligned} \tag{8-29b}$$

A comparison of (8-29a) and (8-29b) shows that they become equivalent in the limit $\omega_1 \to 0$. The two obvious differences in the $\omega_1 \neq 0$ case are the appearance in (8-29a) of a new term

$$\tfrac{1}{2}\omega_1[-(\rho_+^{AB})_{1,1} + (\rho_+^{AB})_{2,2} + (\rho_+^{AB})_{2,3} - (\rho_+^{AB})_{4,1}], \tag{8-29c}$$

which serves to slightly shift the resonance frequency, and the modified interaction with the diagonal elements

$$\tfrac{1}{2}\omega_2[-(\tilde{\rho}_{sr}^{AB})_{1,1} + (\tilde{\rho}_{sr}^{AB})_{2,2} + (\tilde{\rho}_{sr}^{AB})_{2,3}], \tag{8-29d}$$

which has the effect of modifying the strength or amplitude of the resonance as ω_1 is varied.

In double resonance completely new transitions appear which are the $\Delta m = 0$ and $\Delta m = \pm 2$ elements of the density matrix equation. Thus,

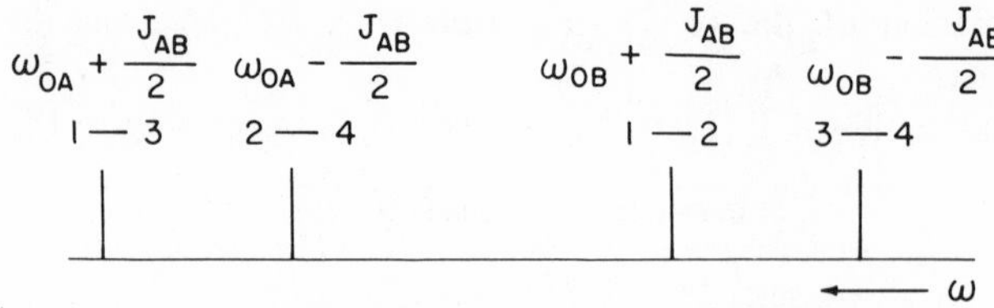

Fig. 8-1. NMR spectrum of the AB system diagrammed at the weak coupling limit.

consider the $\langle 2|\ |3\rangle$ element of (8-20), a $\Delta m = 0$ term,

$$\begin{aligned}0 = i\{&[E_3 - E_2 - (\omega_2 - \omega_1)](\rho_+^{AB})_{2,3} \\ &+ \tfrac{1}{2}\not{\omega}_1[(\rho_+^{AB})_{2,1} - (\rho_+^{AB})_{1,3} + (\rho_+^{AB})_{2,4} - (\rho_+^{AB})_{4,3}] \\ &+ \tfrac{1}{2}J[(\rho_+^{AB})_{2,2} - (\rho_+^{AB})_{3,3}] + \tfrac{1}{2}\not{\omega}_2[(\tilde{\rho}_{sr}^{AB})_{2,4} - (\tilde{\rho}_{sr}^{AB})_{1,3}]\} \\ &- (1/\tau_{AB})(\rho_+^{AB})_{2,3} - [(1/T_{tB}) + (1/T_{lB}) + (1/T_{lA}) + 1/T_{tA}](\rho_+^{AB})_{2,3}\,.\end{aligned} \tag{8-30}$$

A qualitative interpretation of Eq. (8-30) can be obtained if one assumes the weak coupling limit, see Fig. 8-1, which is equivalent to setting equal to zero the term

$$\tfrac{1}{2}J[(\rho_+^{AB})_{2,2} - (\rho_+^{AB})_{3,3}].$$

Let us also neglect for the moment the term in brackets multiplied by $\frac{1}{2}\not{\omega}_1$. Then (8-30) can be written as

$$(\tilde{\rho}_+^{AB})_{2,3} = \frac{\frac{1}{2}i\not{\omega}_2[(\tilde{\rho}_{sr}^{AB})_{2,4} - (\tilde{\rho}_{sr}^{AB})_{1,3}]}{-i[E_3 - E_2 - (\omega_2 - \omega_1)] + 1/T_{eff}}. \tag{8-31}$$

Equation (8-31) will have a resonance when

$$E_3 - E_2 - (\omega_2 - \omega_1) = 0, \tag{8-32}$$

which, substituting from (7-4), becomes

$$\omega_2 = \omega_1 - \omega_{0A} + \omega_{0B}\,. \tag{8-33}$$

As seen from (8-31), the amplitude, i.e., the numerator, of (8-31) will become large when either

$$\omega_1 = E_2 - E_4 = \omega_{0A} - \tfrac{1}{2}J \qquad \text{or} \qquad \omega_1 = E_1 - E_3 = \omega_{0A} + \tfrac{1}{2}J. \tag{8-34}$$

Then substituting (8-34) into (8-33) gives the table

ω_1 at transition	enhanced peak, ω_2
2–4	3–4
1–3	1–2

that is irradiation of the (1–3) transition with ω_1 gives rise to a (1–2) absorption at ω_2.

The matrix element $\langle 3|\ |2\rangle$ will give rise in a similar way to the table

ω_1 at transition	enhanced peak, ω_2
1–2	1–3
3–4	2–4

One must also ask how can $(\rho_+^{AB})_{2,3}$ being large effect the absorption which is given in (8-27) and appears to be independent of $(\rho_+^{AB})_{2,3}$? The answer is that a large $(\rho_+^{AB})_{2,3}$ effects the strength of $(\rho_+^{AB})_{2,1}$, for example, as seen in (8-28), and thus indirectly contributes to the absorption.

A few lines back we dropped the term in Eq. (8-30) multiplied by $\frac{1}{2}\omega_1$. This term serves to shift the $(\rho_+^{AB})_{2,3}$ maximum so it will not fall on top of the direct 1–2 and 3–4 resonances. Thus, we are led to the line splitting seen in Fig. 8.2.

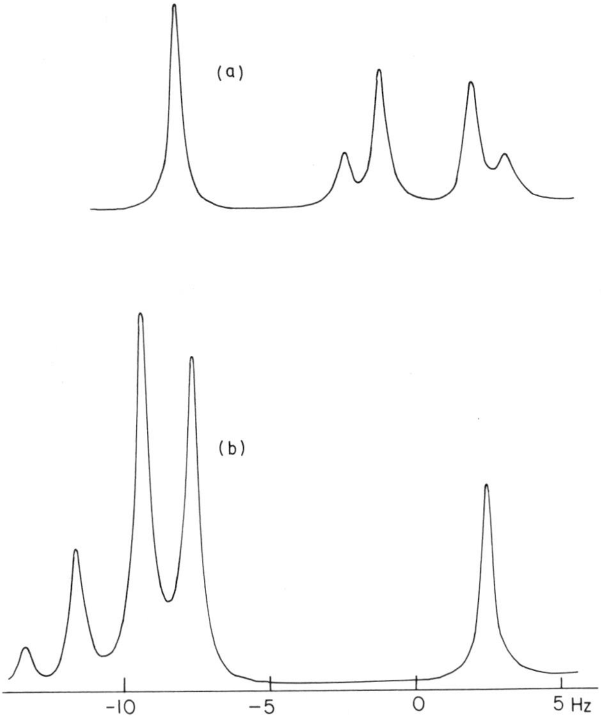

Fig. 8-2. Line splitting in NMDR lineshapes, AB system, $\delta_{AB} = 10$ Hz, $J_{AB} = 4$ Hz, all T_1, T_2 values 12 sec, $\nu_1 = 1.5$ Hz, $\tau = 900$ sec slow exchange limit, (a) $\nu_1 = -12.39$ Hz at $\tilde{\rho}_{1,3}$ transition; (b) $\nu_1 = -1.61$ Hz at $\tilde{\rho}_{1,2}$ transition.

Last, let us consider the $\langle 4|\ |1\rangle$ matrix element of (8-20) which is

$$0 = i\{[E_1 - E_4 - (\omega_2 - \omega_1)](\rho_+^{AB})_{4,1} + \tfrac{1}{2}\omega_1[-(\rho_+^{AB})_{2,1} - (\rho_+^{AB})_{3,1} + (\rho_+^{AB})_{4,2} + (\rho_+^{AB})_{4,3}] + \tfrac{1}{2}\omega_2[(\tilde{\rho}_{sr}^{AB})_{4,2} + (\tilde{\rho}_{sr}^{AB})_{4,3} - (\tilde{\rho}_{sr}^{AB})_{2,1} - (\tilde{\rho}_{sr}^{AB})_{3,1}]\} - [(1/\tau_{AB}) + (1/T_{lA}) + (1/T_{tA}) + (1/T_{lB}) + 1/T_{tB}](\rho_+^{AB})_{4,1} . \tag{8-35}$$

Again, as before, disregarding for the moment the term starting with ω_1 one can write (8-35) as

$$(\rho_+^{AB})_{4,1} = \frac{\tfrac{1}{2}i\omega_2[(\tilde{\rho}_{sr}^{AB})_{4,2} + (\tilde{\rho}_{sr}^{AB})_{4,3} - (\tilde{\rho}_{sr}^{AB})_{2,1} - (\tilde{\rho}_{sr}^{AB})_{3,1}]}{-i[E_1 - E_4 - (\omega_2 - \omega_1)] + 1/T_{eff}} . \tag{8-36}$$

The resonance condition is

$$E_1 - E_4 - \omega_2 - \omega_1 = 0, \tag{8-37}$$

which, after substituting for E_1 and E_4, becomes

$$\omega_2 = \omega_{0A} + \omega_{0B} - \omega_1 . \tag{8-38}$$

As seen from (8-36) the numerator of $(\rho_+^{AB})_{4,1}$ becomes large at each first order resonance, Fig. 8-1. Thus, we can have the following table of frequencies.

ω_1 at transition	enhanced peak, ω_2
3–4	1–3
1–2	2–4
2–4	1–2
1–3	3–4

The preceding discussion serves to give a qualitative understanding of some of the effects seen in double resonance experiments.

For the exact lineshape observed at the set of ω_2 frequencies, it is necessary to solve the full set of 20 coupled ρ_+ equations for the matrix elements listed in Eq. (8-27). The mathematical details as to how this is done have been explained before (3-52). Briefly restated, the procedure is

$$(i(\omega_2 - \omega_1)\mathcal{I} + \mathrm{A})\rho_+ = \mathrm{B} \tag{8-39}$$

and solved as

$$\rho_+ = U(i(\omega_2 - \omega_1)\mathcal{I} + \mathcal{Q})^{-1}U^{-1}\mathrm{B}, \tag{8-40}$$

where $\mathcal{Q}$ is diagonal and

$$\mathrm{A} = U\mathcal{Q}U^{-1}. \tag{8-41}$$

In summary then, in addition to the usual low power absorption (modified as to amplitude) there can be additional fine structure (often not well resolved) given by the following table

ω_1 at transition	enhanced peaks near ω_2
3–4	{1–3, 2–4}
1–2	{2–4, 1–3}
2–4	{1–2, 3–4}
1–2	{2–4, 1–2}

Examples of calculated double resonance lineshapes for the exchanging system (6-200) are shown in the following pages. For comparison, single resonance lineshapes are shown in Fig. 8-3. When B′ is irradiated, effects

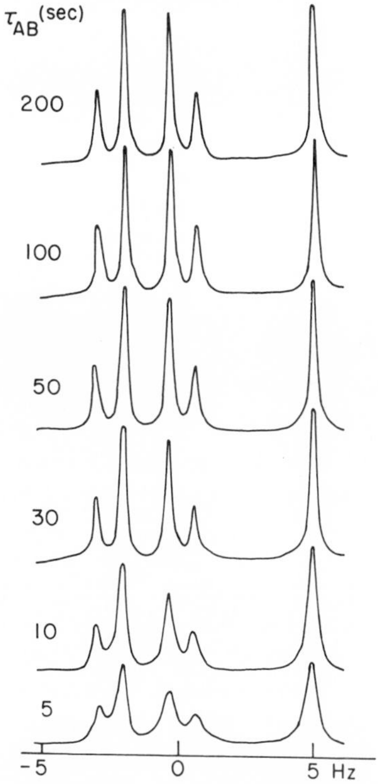

Fig. 8-3. NMR low power single resonance for exchanging system (Fig. 8-1) with (in Hz) ν_A, −2.5; ν_B, 0; $\nu_{B'}$, 5; J_{AB}, 1; all T_1 and T_2 values 10 sec, (AB) = (B′), τ_{AB} values noted on curves.

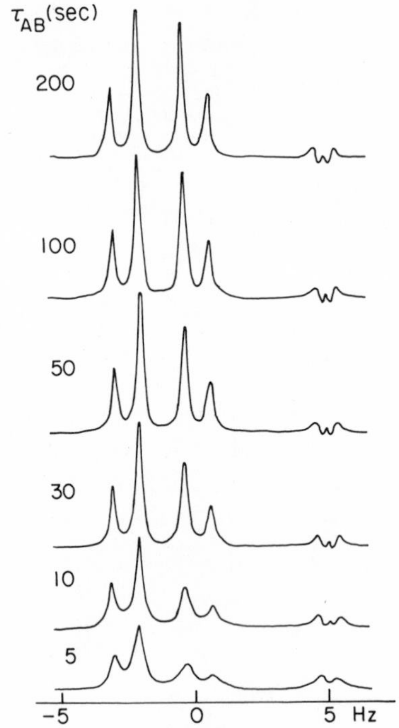

Fig. 8-4. NMDR lineshapes for exchanging system (Fig. 8-1) with irradiation of the B′ resonance, $\nu_1 = 0.3$ Hz and $\nu_1 = 5$ Hz, other parameters as in Fig. 8-3; τ_{AB} values noted on curves.

in the AB lineshape are seen due to transfer of saturation by exchange, see Fig. 8-4, the steady state equivalent of a Hoffmann–Forsén experiment.[7, 8] Lineshapes which result from strongly irradiating B′ are far more sensitive to slow rates of exchange than those obtained at low power, compare Fig. 8-3 with Fig. 8-4.

The shapes of the secondary splittings and distortions induced by strongly irradiating the AB transitions (singly) are also effected by exchange, see Figs. 8-5 and 8-6. Both the secondary splittings and scalar coupling are seen to average as a result of exchange. The extra fine structure causes these lineshapes also to be more sensitive to rate changes than at low power. However, this is seen most clearly over the range $\tau_{AB} = 10$–1 sec and not for slow exchange rates. A double resonance procedure which gives information on even slower rates of exchange has been proposed.[14]

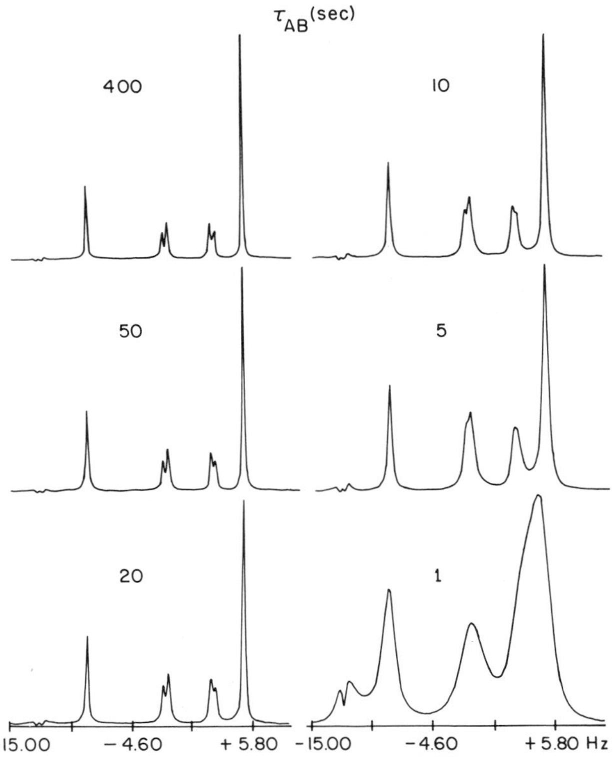

Fig. 8-5. NMDR lineshapes for exchanging system (Fig. 8-1) with parameters on Hz, ν_{0A}, -10; ν_{0B}, 0; $\nu_{0B'}$, $= +5$; J_{AB}, $+4$; all T_1 and T_2 values 12 sec, ν_1 at -12.39 Hz, the 1←3 transition and $\not{\nu}_1 =$ 0.5 Hz; (AB) = (B), $\tau_{AB} = \tau_{B'}$ values on figure.

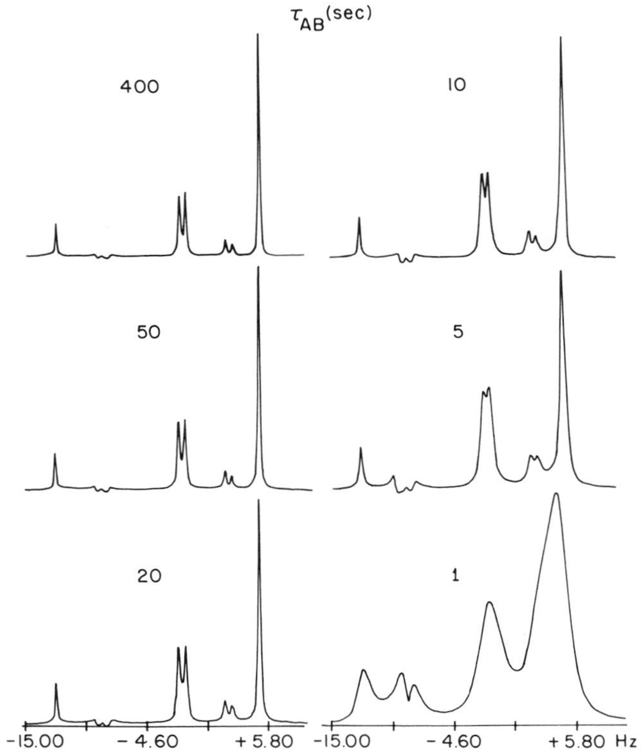

Fig. 8-6. NMDR lineshapes, system (Fig. 8-1). All parameters as for Fig. 8-5 except $\nu_1 = -8.39$ Hz, the 2←4 transition.

2. Overhauser Effects

Overhauser effects seen in double resonance experiments arise from intramolecular dipolar relaxation (see Chapter IV). The geometrical dependence of this interaction can be utilized as an aid in structure determination. All that is needed is to include this operator in the equations for ρ_+ and ρ. Having done this, one would be able to study effects on NMR lineshapes of exchanging systems subject to intramolecular dipolar relaxation.

3. Experimental NMDR Lineshapes in Exchanging Systems

Yang and Gordon[15] reported double resonance proton lineshapes for 2,2,2-trichloroethanol in carbon disulfide. They calculated the lineshapes using the density matrix treatment as a function of the exchange rate for the process

$$CCl_3CH_2O^*H^* + CCl_3CH_2OH \rightleftharpoons CCl_3CH_2OH^* + CCl_3CH_2O^*H \qquad (8\text{-}42)$$

and the power of the irradiating field. The lineshapes depended significantly on these two parameters. Extensive transfer of saturation by exchange was noted, see Figs. 8-7 and 8-8. However, no kinetic analysis of exchange was made. Relaxation was considered to arise from the uncorrelated random field mechanism. Also, an inhomogeneity term was added to diagonal elements of the coefficient matrix.[§] Typical results are shown in Figs. 8-7 and 8-8. It was noted that these double resonance lineshapes were more sensitive to rate changes for slow exchange than is the case for single resonance.

In related work Fung and Olympia[16] investigated rotation in 1-chloro-2,2-difluoro-1,1,2-tribromoethane. The fluorines in the symmetrical rotamer, **1**

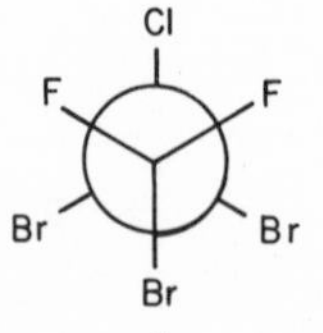

1

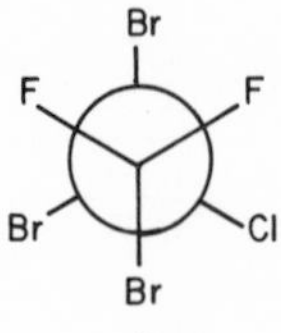

2

are equivalent, while those in **2** give an AB system. Interconversion of rotamers brings about signal averaging. Low rf single resonance was utilized to obtain the barrier to rotation. Density matrix equations for the

[§]This procedure only applies correctly at low power.

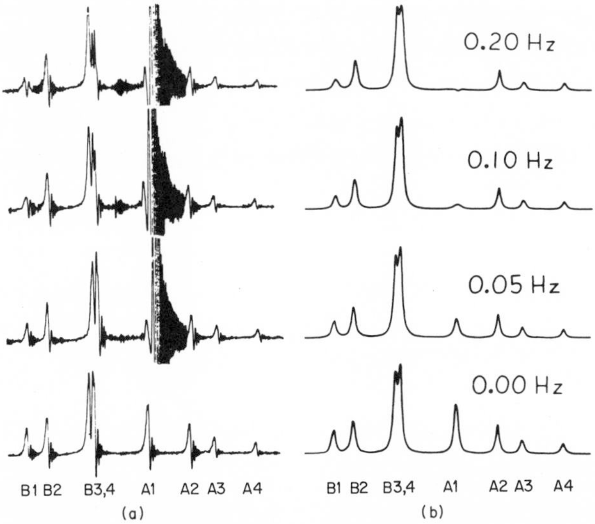

Fig. 8-7. Experimental (a) and theoretical (b) NMDR lineshapes for 2,2,2-trichloroethanol undergoing intermolecular OH exchange with irradiation of peak Al at different rf fields, labeled. Theoretical spectra are computed for $\tau = 4$ sec., $T_A = T_B = 88$ sec, and λ (inhomogeneity) = 0.2 Hz. [From Ref. 14.]

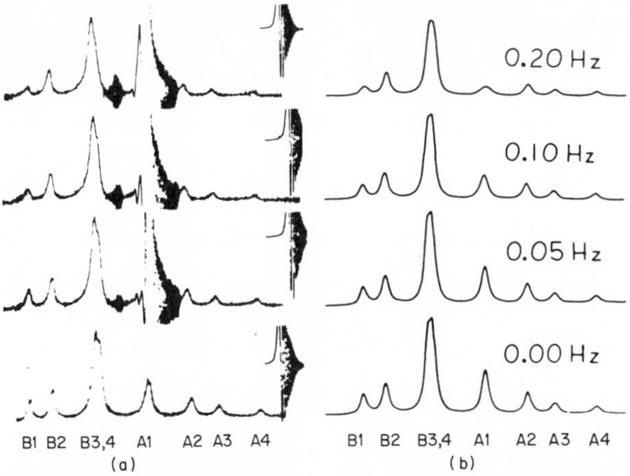

Fig. 8-8. NMDR experimental (a) and theoretical (b) spectra as in Fig. 8-7. with $\tau = 0.5$ sec, and $T_A = T_B = 7$ sec. [From Ref. 14.]

double resonance system were derived. These were not utilized for analysis of the rotation process. Here, also, the intermolecular random field interaction was considered to be the significant relaxation mechanism.

Appendix A. Double Resonance Hamiltonian in the Rotating Frame

The following treatment is for a single spin but can easily be generalized for several.

Given ρ in the laboratory frame related to $\tilde{\rho}$ in the rotating frame by

$$\rho = e^{-i\omega_1 t I^z}\tilde{\rho}e^{+i\omega_1 t I^z}, \tag{8-A1}$$

the density matrix equation in the rotating coordinate system becomes

$$\dot{\tilde{\rho}} = -i\left[\mathcal{H}_0 - \omega_1 I^z, \tilde{\rho}\right] - i\left[\overline{\mathcal{H}}_{\mathrm{rf1}}, \tilde{\rho}\right] - i\left[\overline{\mathcal{H}}_{\mathrm{rf2}}, \tilde{\rho}\right] + R\tilde{\rho} + E\tilde{\rho}. \tag{8-A2}$$

Next, we evaluate the Hamiltonians in the rotating systems as

$$\overline{\mathcal{H}}_{\mathrm{rf1}} = e^{+i\omega_1 t I^z}\mathcal{H}_{\mathrm{rf1}}e^{-i\omega_1 t I^z}, \tag{8-A3}$$

$$\mathcal{H}_{\mathrm{rf1}} = \tfrac{1}{2}\omega_1\left[(I^+ + I^-)\cos\omega_1 t - i(I^+ - I^-)\sin\omega_1 t\right]$$

$$= \tfrac{1}{2}\omega_1\left[I^+ e^{-i\omega_1 t} + I^- e^{+i\omega_1 t}\right], \tag{8-A4}$$

$$\overline{\mathcal{H}}_{\mathrm{rf1}} = \tfrac{1}{2}\omega_1\left[e^{+i\omega_1 t I^z}I^+ e^{-i\omega_1 t I^z}e^{-i\omega_1 t} + e^{+i\omega_1 t I^z}I^- e^{-i\omega_1 t I^z}e^{+i\omega_1 t}\right], \tag{8-A5}$$

$$= \tfrac{1}{2}\omega_1\left[I^+ + I^-\right] = \omega_1 I^x. \tag{8-A6}$$

Transformation (8-A5) to (8-A6) is described following (3-17). The Hamiltonian for the weaker driving field in the rotating frame is

$$\overline{\mathcal{H}}_{\mathrm{rf2}} = e^{i\omega_1 t I^z}\mathcal{H}_{\mathrm{rf2}}e^{-i\omega_1 t I^z}, \tag{8-A7}$$

where

$$\mathcal{H}_{\mathrm{rf2}} = \tfrac{1}{2}\omega_2\left[I^+ e^{-i\omega_2 t} + I^- e^{i\omega_2 t}\right]. \tag{8-A8}$$

Equation (8-A7) is evaluated in a similar manner to (8-A3). The result is

$$\overline{\mathcal{H}}_{\mathrm{rf2}} = \tfrac{1}{2}\omega_2\left[I^+ e^{i(\omega_1 - \omega_2)} + I^- e^{-i(\omega_1 - \omega_2)}\right], \tag{8-A9}$$

$$= \tfrac{1}{2}\omega_2\left[I^+(\cos\Delta\omega t + i\sin\Delta\omega t) + I^-(\cos\Delta\omega t - i\sin\Delta\omega t)\right] \tag{8-A10}$$

$$= \tfrac{1}{2}\omega_2\left[I^x\cos\Delta\omega t - I^y\sin\Delta\omega t\right], \tag{8-A11}$$

where

$$\Delta\omega = \omega_1 - \omega_2 . \tag{8-A12}$$

Appendix B. Glossary of Symbols

Abs	Absorption
a, b, c	Labels of product basis functions
E	Exchange operator
N	Number of spin states
R	Relaxation operator
T_{1i}, T_{2i}	Relaxation times, nucleus i
$\mathcal{H}$	Hamiltonian, laboratory frame
$\bar{\mathcal{H}}$	Hamiltonian, rotating frame
ν_{0i}	chemical shift nucleus i (Hz)
ν_1, ν_2	rf frequencies (Hz)
$\not{\nu}_1, \not{\nu}_2$	rf power (Hz)
ρ	Density matrix, laboratory frame
$\tilde{\rho}$	Density matrix, rotating frame
$\tilde{\rho}_{sr}$	Density matrix rotating frame, single resonance, all rf fields
$\rho_{\pm}$	perturbation term due to second rf field
ω_{0i}	chemical shift nucleus i (rad/sec)
ω_1, ω_2	rf frequencies (rad/sec)
$\not{\omega}_1, \not{\omega}_2$	rf power (rad/sec)

Problem

1. Consider two exchanging sites A and B with concentrations (A) and (B). If one saturates B with $\not{\omega}_1$, find M_z^{A} assuming that the chemical shifts are large enough so that the saturating field has no direct effect on A. If one applies a low power field $\not{\omega}_2$ near $\omega_{0\mathrm{A}}$, while $\not{\omega}_1$ is still on, find the absorption of A. Show that the relative absorption of A with and without the saturating field $\not{\omega}_1$ is given as

$$\frac{\mathrm{Abs}(\not{\omega}_1)}{\mathrm{Abs}(0)} = \frac{\left(T_{1\mathrm{A}}^{-1} + \tau_{\mathrm{AB}}^{-1}\right)^{-1}}{T_{1\mathrm{A}}}.$$

REFERENCES

1. W. A. Anderson, *Phys. Rev.* **102**, 151 (1956).
2. J. D. Baldeschwieler, *J. Chem. Phys.* **40**, 459 (1964).
3. B. D. Nageswara Rao, *Phys. Rev. A* **137**, 467 (1965).
4. B. D. Nageswara Rao, J. D. Baldeschwieler, and J. M. Anderson, *Phys. Rev. A* **137**, 1477 (1965).
5. J. D. Baldeschwieler and E. W. Randall, *Chem. Rev.* **63**, 81 (1963).
6. W. von Phillipsborn, *Angew. Chem., Int. Ed. Engl.* **10**, 472 (1963).
7. S. Forsen and R. A. Hoffmann, *J. Chem. Phys.* **40**, 1189 (1964).

8. R. A. Hoffmann and S. Forsen, *J. Chem. Phys.* **45**, 2049 (1966).
9. J. H. Noggle and R. E. Schirmer, "The Nuclear Overhauser Effect: Chemical Applications," pp. 126–165. Academic Press, New York, 1971.
10. D. S. Kavalokoff and E. Namanwirth, *J. Amer. Chem. Soc.* **92**, 3234 (1970).
11. P. W. N. M. Van Leenwen and H. P. Praat, *J. Organomet. Chem.* **22**, 483 (1970).
12. J. I. Kaplan, P. P. Yang, and G. Fraenkel, *J. Amer. Chem. Soc.* **97**, 3881 (1975).
13. J. M. Anderson, *J. Magn. Reson.* **4**, 184 (1971).
14. J. I. Kaplan and R. E. Carter, *J. Magn. Reson.* **33**, 437 (1979).
15. P. P. Yang and S. L. Gordon, *J. Chem. Phys.* **54**, 1779 (1971).
16. B. M. Fung and P. M. Olympia, *Mol. Phys.* **19**, 685 (1970).

Chapter IX

TRANSIENT EFFECTS

1. Introduction

So far in this book we have only considered continuous wave (cw) responses. In this chapter we will discuss two kinds of transient responses. One will be the so-called free induction decay (FID) and in particular the Fourier transform relationship between the FID and the cw low power absorption. The other effect will be the observation of the decay of a saturated cw absorption by use of a low power signal at a different frequency.[1, 2]

Before going on, we must make the following proviso: we are assuming that the relaxation operator derived for steady state conditions is still appropriate for transient responses.

2. Fourier Relation of the Free Induction Decay to the Low Power Absorption

First, let us calculate the low power absorption in a new manner. We start with the general density matrix equation

$$\dot{\rho} = -i[\mathcal{H}_0 + \mathcal{H}_{\text{rf}}, \rho] + R\rho + E\rho, \tag{9-1}$$

where E means the exchange terms are linearized, see Chapter V, and

$$\mathcal{H}_{\text{rf}} = \gamma\!\!\!/_1 I^x \cos \omega t. \tag{9-2}$$

Next, rewrite Eq. (9-1) as

$$\dot{\rho} = (\mathcal{L}_s + \mathcal{L}_R + \mathcal{L}_E)\rho + \mathcal{L}_{\text{rf}}\rho \cos \omega t, \tag{9-3}$$

where

$$\mathcal{L}_s\rho = -i[\mathcal{H}_S, \rho], \tag{9-4a}$$

$$\mathcal{L}_R\rho = R\rho, \tag{9-4b}$$

$$\mathcal{L}_E\rho = E\rho, \tag{9-4c}$$

$$\mathcal{L}_{\text{rf}}\rho = -i[\mathcal{H}_{\text{rf}}, \rho], \tag{9-4d}$$

and follow this by going into the interaction representation by the transformation

$$\rho = e^{(\mathcal{L}_s + \mathcal{L}_R + \mathcal{L}_E)t}\tilde{\rho}. \tag{9-5}$$

Substituting (9-5) into (9-3) one obtains

$$\dot{\tilde{\rho}} = e^{-(\mathcal{L}_s + \mathcal{L}_R + \mathcal{L}_E)t}\mathcal{L}_{\text{rf}}\rho(t) \cos\omega t \tag{9-6}$$

or

$$\tilde{\rho}(t) = \int_{-\infty}^{t} e^{-(\mathcal{L}_s + \mathcal{L}_R + \mathcal{L}_E)t'}\mathcal{L}_{\text{rf}} (\cos \omega t')\rho(t')\, dt', \tag{9-7}$$

$$\rho(t) = \int_{-\infty}^{t} e^{(\mathcal{L}_s + \mathcal{L}_R + \mathcal{L}_E)(t-t')}\mathcal{L}_{\text{rf}}(\cos \omega t')\rho(t')\, dt'. \tag{9-8}$$

Making the change of variables,

$$t - t' = u, \tag{9-9}$$

(9-8) becomes

$$\rho(t) = \int_{0}^{\infty} e^{(\mathcal{L}_s + \mathcal{L}_R + \mathcal{L}_E)u}\mathcal{L}_{\text{rf}} (\cos \omega(t-u))\rho(t-u)\, du. \tag{9-10}$$

As we want $\rho(t)$ only to first order in ω_1 (low power or linear response), we can replace $\rho(t-u)$ on the right-hand side of (9-10) by its equilibrium value ρ_0. Additionally we note that

$$\cos \omega(t-u) = \cos \omega t \cos \omega u + \sin \omega t \sin \omega u.$$

Thus, the absorption, which is proportional to the component of magnetization out of phase with the rf field, is given as

$$\text{Abs} \simeq \text{Tr} \int_{0}^{\infty} I^x e^{(\mathcal{L}_s + \mathcal{L}_R + \mathcal{L}_E)u}\mathcal{L}_{\text{rf}} (\sin \omega u)\rho_0\, du. \tag{9-11}$$

Substituting for the standard high temperature approximation for ρ_0

$$\rho_0 \simeq (\mathcal{I} - \hbar\omega_0 I^z/kT_b)/N \tag{9-12}$$

into (9-4c) one obtains

$$\mathcal{L}_{\text{rf}}\rho_0 = -i\hbar\omega_0 I^y/NkT_b, \tag{9-13}$$

which substituted into (9-11) gives as our final result that

$$\text{Abs} \propto \int_0^\infty I^x e^{(\mathcal{L}_s + \mathcal{L}_R + \mathcal{L}_E)u} I^y (\sin \omega u)\, du. \tag{9-14}$$

Next, consider the free induction decay. We start with the equilibrium distribution ρ_0 given in (9-12). The system is rotated by a 90° pulse about the x axis so that

$$\rho_0' = e^{i(\pi/2)I^x} \rho_0 e^{-i(\pi/2)I^x} = (\mathcal{I} - \hbar\omega_0 I^y / kT_b)/N \tag{9-15}$$

and then the system is allowed to decay. The system we wish to solve is

$$\dot{\rho} = (\mathcal{L}_s + \mathcal{L}_R + \mathcal{L}_E)\rho \tag{9-16}$$

with the initial condition given by (9-15). The solution thus is given as

$$\rho(t) = e^{(\mathcal{L}_s + \mathcal{L}_R + \mathcal{L}_E)t} \rho_0', \tag{9-17}$$

which gives rise to the x component of magnetization of the free induction decay

$$M_x \propto -(\hbar\omega_0 / kT_b) \operatorname{Tr} I^x e^{(\mathcal{L}_s + \mathcal{L}_R + \mathcal{L}_E)t} I^y, \tag{9-18}$$

where it seen that Eq. (9-14) is the Fourier Transform of (9-18).

3. Saturation Decay in an Exchanging System Detected by a Small rf Field

Consider the first order exchanging system

$$\mathrm{A} \rightleftharpoons \mathrm{B},$$

where A and B each contain one spin or a group of equivalent spins.

Let us begin by looking at this problem from the point of view of the Bloch equations

$$\dot{M}_z^{\mathrm{A}} = -\frac{1}{T_{1\mathrm{A}}}(M_z^{\mathrm{A}} - M_z^{0\mathrm{A}}) - \frac{1}{\tau_{\mathrm{A}}} M_z^{\mathrm{A}} + \frac{1}{\tau_{\mathrm{B}}} M_z^{\mathrm{B}}, \tag{9-19}$$

$$\dot{M}_z^{\mathrm{B}} = -\frac{1}{T_{1\mathrm{B}}}(M_z^{\mathrm{B}} - M_z^{0\mathrm{B}}) - \frac{1}{\tau_{\mathrm{B}}} M_z^{\mathrm{B}} + \frac{1}{\tau_{\mathrm{A}}} M_z^{\mathrm{A}}. \tag{9-20}$$

We irradiate the B system with a strong enough rf field so that

$$M_z^{\mathrm{B}} = 0. \tag{9-21}$$

Then, from (9-19), we obtain for the steady state solutions, $\dot{M}_z^{\mathrm{A}} = 0$,

$$M_{\mathrm{A}}^z = M_z^{0\mathrm{A}} / (1 + T_1/\tau_{\mathrm{A}}). \tag{9-22}$$

Equations (9-21) and (9-22) will be our initial conditions for the transient solution.

We now imagine that a low power rf field is turned on at the resonance frequency of the A system. We will equate the magnitude of the absorption of the A system to be proportional to M_z^A, as shown in (1-17). Thus, we need to solve (9-19) and (9-20) with initial condition (9-21) and (9-22). The solutions will be of the form

$$M_z^A(t) = M_z^{0A} + M_A^1 e^{-\alpha_1 t} + M_A^2 e^{-\alpha_2 t}, \tag{9-23}$$

$$M_z^B(t) = M_z^{0B} + M_B^1 e^{-\alpha_1 t} + M_B^2 e^{-\alpha_2 t}. \tag{9-24}$$

Substituting (9-23) and (9-24) into (9-19) and (9-20), respectively, one obtains

$$\begin{bmatrix} (1/T_{1A}) + 1/\tau_A & -1/\tau_B \\ -1/\tau_A & (1/T_{1B}) + 1/\tau_B \end{bmatrix} \begin{bmatrix} M_z^{0A} \\ M_z^{0B} \end{bmatrix} = \begin{bmatrix} M_z^{0A}/T_{1A} \\ M_z^{0B}/T_{1B} \end{bmatrix} \tag{9-25}$$

and

$$\begin{bmatrix} (1/T_{1A}) + (1/\tau_B) - \alpha_1 & -1/\tau_B \\ -1/\tau_A & (1/T_{1B}) + (1/\tau_B) - \alpha_1 \end{bmatrix} \begin{bmatrix} M_A^1 \\ M_B^1 \end{bmatrix} = 0. \tag{9-26}$$

The α_1 and α_2 values are thus obtained by setting the determinant of the 2×2 matrix given in (8-26) equal to zero or

$$\left(\frac{1}{T_{1A}} + \frac{1}{\tau_A} - \alpha_1\right)\left(\frac{1}{T_{1B}} + \frac{1}{\tau_B} - \alpha_1\right) - \frac{1}{\tau_A}\frac{1}{\tau_B} = 0 \tag{9-27}$$

$$\alpha_1^2 - \alpha_1\left(\frac{1}{T_{1A}} + \frac{1}{T_{1B}} + \frac{1}{\tau_A} + \frac{1}{\tau_B}\right) + \frac{1}{T_{1A}T_{1B}} + \frac{1}{T_{1A}\tau_B} + \frac{1}{T_{1B}\tau_A} = 0 \tag{9-28}$$

with the solutions for the special case that

$$\begin{aligned} T_{1B} &= T_{1A} \quad \text{and} \quad \tau_A = \tau_B, \\ \alpha_1 &= (1/T) + 2/\tau, \quad \alpha_2 = 1/T. \end{aligned} \tag{9-29}$$

The experimenter can thus obtain information on τ by monitoring the decay of the magnetization. In what follows we will solve the same sort of problem but generalize it to an arbitrary system using the density matrix formalism.

We could start off by writing down the density matrix equations for some particular system undergoing exchange, for instance (8-1). Instead, to keep the problem general, we write

$$\dot{\rho} = -i[\mathcal{H}_s + \mathcal{H}_{rf1}, \rho] + R\rho + E\rho \tag{9-30}$$

to cover any set of chemically exchanging systems. Going into the rotating coordinate system (9-30) becomes in the steady state

$$0 = -i[\overline{\mathcal{H}}_s + \overline{\mathcal{H}}_{rf1}, \tilde{\rho}'] + R\tilde{\rho}' + E\tilde{\rho}'. \tag{9-31}$$

The solution of (9-31) has been described in Chapter VII. At time $t = 0$ we turn off the large rf signal, ω_1, and turn on the small observing field, ω_2. The density matrix equation in the rotating coordinate system at frequency ω_1 is given for $t > 0$ as

$$\dot{\tilde{\rho}} = -i[\overline{\mathcal{H}}_s + \overline{\mathcal{H}}_{rf2}, \tilde{\rho}] + R\tilde{\rho} + E\tilde{\rho}, \tag{9-32}$$

where

$$\overline{\mathcal{H}}_{rf2} = \tfrac{1}{2}\omega_2[I^+e^{-i(\omega_2-\omega_1)t} + I^-e^{i(\omega_2-\omega_1)t}]. \tag{9-33}$$

The problem then is to solve (9-32) to first order in ω_2 with the initial condition given by the solution of (9-31). Thus, we look for a solution in the form

$$\tilde{\rho}(t) = \tilde{\rho}_0(t) + [\rho_+(t)e^{-i(\omega_2-\omega_1)t} + \rho_-(t)e^{i(\omega_2-\omega_1)t}], \tag{9-34}$$

where ρ^+ and ρ^- are linear with respect to ω_2 and are related by

$$\rho_+ = (\rho^-)^\dagger$$

(hermitian). Substituting (9-34) into (9-32) and collecting terms with the same Fourier components, one obtains the homogeneous equations

$$\dot{g} = -i[\overline{\mathcal{H}}_s, g] + R'g + Eg \tag{9-35}$$

with the initial condition that at $t = 0$[§]

$$g(0) = \tilde{\rho}' - \rho_0 \tag{9-36}$$

and the inhomogeneous equations

$$\dot{\rho}_+ = i(\omega_2 - \omega_1)\rho_+ - i[\overline{\mathcal{H}}_s, \rho_+] - i[\overline{\mathcal{H}}_{rf}, \rho_0 + g] + R'\rho_+ + E\rho_+, \tag{9-37}$$

where

$$\overline{\mathcal{H}}_{rf2} = \tfrac{1}{2}\omega_2 I^+ \tag{9-38}$$

and R' means the relaxation operator operates on g or ρ_+ and not on $g - \rho_0$ or $\rho_+ - \rho_0$.

[§]The actual initial condition is that $\rho_0 + g(0) + \rho_+(0) + \rho_-(0) = \tilde{\rho}'$. This correct boundary condition will give rise to higher order terms in ω_2 and can be neglected.

Taking matrix elements of (9-35) in the product representation, one obtains (9-35) in the form

$$\dot{g}(t)_{\text{col}} = Ag(t)_{\text{col}} \tag{9-39}$$

Equation (9-39) with the boundary condition (9-36) is solved as

$$g(t)_{\text{col}} = e^{At} g(0)_{\text{col}}. \tag{9-40}$$

Next, we write

$$\text{A} = U_{\text{A}} \text{A}_{\text{d}} U_{\text{A}}^{-1}, \tag{9-41}$$

where A_{d} is a diagonal matrix and the evaluation of U is described in Chapter VI.

Substituting (9-41) into (9-40), one obtains

$$g(t)_{\text{col}} = e^{U_{\text{A}}\text{A}_{\text{d}}U_{\text{A}}^{-1}t} g(0)_{\text{col}} = U_{\text{A}} e^{\text{A}_{\text{d}}t} U_{\text{A}}^{-1} g(0)_{\text{col}}. \tag{9-42}$$

Proof:

$$\begin{aligned} e^{U_{\text{A}}\text{A}_{\text{d}}U_{\text{A}}^{-1}t} &= 1 + U_{\text{A}}\text{A}_{\text{d}}U^{-1}t + (1/2!)U_{\text{A}}\text{A}_{\text{d}}U_{\text{A}}^{-1}U_{\text{A}}\text{A}_{\text{d}}U_{\text{A}}^{-1}t^2 \\ &= U_{\text{A}}\{1\}U_{\text{A}}^{-1} + U_{\text{A}}\{\text{A}_{\text{d}}t\}U_{\text{A}}^{-1} + U_{\text{A}}\{(1/2!)(\text{A}_{\text{d}}t)^2\}U_{\text{A}}^{-1} \end{aligned} \tag{9-43}$$

$$\begin{aligned} &= U_{\text{A}}\left[1 + \text{A}_{\text{d}} + (1/2!)\text{A}_{\text{d}}^2\right]U_{\text{A}}^{-1} \\ &= U_{\text{A}} e^{\text{A}_{\text{d}}t} U_{\text{A}}^{-1}, \end{aligned} \tag{9-44}$$

where =

$$e^{\text{A}_{\text{d}}t} = \begin{bmatrix} e^{\text{A}_{11}t} & & 0 \\ & e^{\text{A}_{22}t} & \\ & & \ddots \\ 0 & & e^{\text{A}_{nn}t} \end{bmatrix} \tag{9-45}$$

Substituting for $\tilde{\rho}_0 + g$ into (9-37) and taking matrix elements of (9-37) in the product representation, one has

$$\dot{\rho}_{+\text{col}} = \text{B}\rho_{+\text{col}} + D(t)_{\text{col}}, \tag{9-46}$$

where $D(t)_{\text{col}}$ are the product representation matrix elements of

$$-i\left[\overline{\mathcal{H}}_{\text{rf}}, \rho_0 + g(t)\right].$$

The inhomogeneous solution of (9-46) is given as

$$\rho_+(t)_{\text{col}} = \int_0^t e^{\text{B}(t-t')} D(t')\, dt', \tag{9-47}$$

$$\rho_+(t)_{\text{col}} = U_{\text{B}} \int_0^t e^{\text{B}_{\text{d}}(t-t')} U_{\text{B}}^{-1} D(t')\, dt', \tag{9-48}$$

where B_d is a diagonal matrix. The integrand of Eq. (9-48) will be of the form

$$\begin{bmatrix} e^{B_{11}(t-t')} & & \\ & \ddots & \\ & & e^{B_{nn}(t-t')} \end{bmatrix} \begin{bmatrix} U_{B_{11}}^{-1} & U_{B_{12}}^{-1} \cdots \\ U_{B_{21}}^{-1} & \cdots \\ U_{\hat{B}_{\hat{n}1}}^{-1} & \end{bmatrix} \begin{bmatrix} \sum_{n'} D_1^{n'} e^{A_{n'n'}t'} + C_1 \\ \vdots \\ \sum_{n'} D_n^{n'} e^{A_{n'n'}t'} + C_n \end{bmatrix}. \tag{9-49}$$

The integrations are straightforward as they will all be of the form

$$\int_0^t e^{B_{mm}(t-t')} e^{A_{11}t'} = \frac{e^{B_{mm}t}\left[e^{(A_{11}-B_{mm})t} - 1\right]}{A_{11} - B_{mm}}. \tag{9-50}$$

The x component of magnetization in the laboratory frame is as

$$M_x(t) = \mathrm{Tr}\, I^x \rho(t). \tag{9-51}$$

Going to the rotating frame at frequency ω_1 (9-51) becomes (see Chapter VIII)

$$\begin{aligned} M_x(t) &= \mathrm{Tr}\, \tfrac{1}{2}\left[I^+ e^{i\omega_1 t} + Ie^{-i\omega_1 t}\right]\tilde{\rho}(t) \\ &= \tfrac{1}{2}\, \mathrm{Tr}\left[I^+ e^{i\omega_1 t} + I^- e^{-i\omega_1 t}\right]\left[\rho_0 + g + \rho_+ e^{-i(\omega_2-\omega_1)t} + \rho_- e^{i(\omega_2-\omega_1)t}\right]. \end{aligned} \tag{9-52}$$

The response we are looking for is the term in Eq. (9-52) which is multiplied by $\sin \omega_2 t$ (Chapter VIII), see Problem 4.

One might ask why do we so want such a complicated procedure? Its main advantage is that for slow exchange the small changes arising from exchange are made time dependent and thus easier to observe. An analysis of the observed signal as a means of determining the exchange rate for all but the simplest systems, as this chapter shows, involves considerable computer programming.

Problems

1. Calculate the effect of truncating the integral $\int_0^\infty$ (i.e., replace it by $\int_0^T$) on the lineshape obtained by Fourier transforming the free induction decay of a single spin.

2. From the free induction decay for noninteracting spins, calculate the absorption using (9-15).

3. Consider two isolated species, each spin $\frac{1}{2}$, called A and B, which exchange environments as a result of some chemical process. Their Bloch equations for the M_z components alone will be

$$\frac{dM_z^A}{dt} = -\frac{(M_z^A - M_z^{0A})}{T_{1A}} - \frac{1}{\tau_A}(M_z^A - M_z^B), \qquad \text{(9-P1)}$$

$$\frac{dM_z^B}{dt} = -\frac{(M_z^B - M_z^{0B})}{T_{1B}} - \frac{1}{\tau_B}(M_z^B - M_z^A). \qquad \text{(9-P2)}$$

(a) For steady state solution, calculate M_z^A when a large rf field acts on B.

(b) Use the previous result as the initial condition for the solution of the coupled equations (9-P1) and (9-P2).

4. Find the term in (9-52) which goes as $\sin \omega_2 t$.

5. Consider a two site exchanging system. The sites are occupied by spin $\frac{1}{2}$ species called A and B and in concentrations (A) and (B) and with relaxation times T_{1A} and T_{1B}. If one inverts the populations with a 180° pulse, calculate the system relaxation times.

REFERENCES

1. S. Forsen and R. A. Hoffmann, *J. Chem. Phys.* **40**, 1189 (1964).
2. R. A. Hoffmann and S. Forsen, *J. Chem. Phys.* **45**, 2049 (1966).

INDEX